OTHER TITLES OF INTEREST FROM ST. LUCIE PRESS

Resolving Environmental Conflict: Towards Sustainable Community Development

Development, Environment, and Global Dysfunction: Toward Sustainable Recovery

Sustainable Community Development

Environmental Management Tools on the Internet: Accessing the World of Environmental Information

Economic Theory for Environmentalists

Sustainable Forestry: Philosophy, Science and Economics

Ecology and the Biosphere

For more information about these titles call, fax or write:

St. Lucie Press
100 E. Linton Blvd., Suite 403B
Delray Beach, FL 33483
TEL (561) 274-9906 • FAX (561) 274-9927
E-MAIL: information@slpress.com
WEB SITE: http://www.slpress.com

S^{t}L

Principles of Sustainable Development

Principles of

Sustainable

Development

Edited by
F. Douglas
Muschett

C. Lee Campbell

Victoria Evans

Walter W. Heck

Si Duk Lee

Victor S. Lee

F. Douglas Muschett

Thomas T. Shen

John L. Warren

S^t_L

St. Lucie Press
Delray Beach, Florida

Phone: (561) 274-9906
Fax: (561) 274-9927
E-mail: information@slpress.com
Web site: http://www.slpress.com

S$_L^t$

Published by
St. Lucie Press
100 E. Linton Blvd., Suite 403B
Delray Beach, FL 33483

CONTENTS

PREFACE

The concerns related to sustainable development have been present for many years. However, it is only within the past few years that sustainable development has assumed prominence as an important concept and philosophy to guide economic development and environmental management. Consequently, there is still much confusion about sustainable development—even among environmental professionals. Frequently there is a tendency to view and present sustainable development in rather narrow terms, such as "new environmental technologies" or "population stabilization."

And yet, if the developed and developing countries are to adopt sustainable development as a central organizing principle— as they must to maintain life support systems, secure a healthy environment and promote widespread prosperity—a multifaceted approach is necessary. This book brings together the collective thinking, perspectives and experiences of several individuals from different disciplines who are working to advance and operationalize sustainable development. The book was inspired by a provocative panel discussion which was held at the Annual Meeting of the Air and Waste Management Association in Cincinnati in June 1994.

In the opening chapter I present an overview and integrated approach to sustainable development. Using historical and social contexts, I discuss how sustainable development is vital to the prosperity of civilizations. Using a systems approach and two conceptual models, I demonstrate how sustainable development can be advanced; economic growth, which is less resource intensive and less polluting, can be effected through better integration of natural resource and environmental management by using several types of economic incentives. I stress the importance of interdisciplinary thinking and analysis for discovering "win–win" situations which benefit both economic development and the environment. Finally, I argue that ambient environmental management must be better integrated and include more creative strategies, including land use practices, which will yield multiple environmental benefits.

C. Lee Campbell and Walter Heck analyze different ecological perspectives on sustainability and present them within a balanced framework of ecological, social and economic well-being. In particular, they emphasize that ecological sustainability must occur in conjunction with land use management at several different geographic scales—ranging from global to regional to landscapes—where traditionally there has been little ecological management and little data collection. The authors conclude their chapter with a description of the new Environmental Monitoring and Assessment Program, which is intended to provide ecological indicators over time and at geographic scales to provide needed information for ecological management.

Si Duk Lee and Victor Lee review the global concern among the international community for sustainable development. Unfortunately, they conclude that since the 1992 Conference on Environment and Development in Rio de Janeiro, there has been little international agreement; they discuss how the developed and developing countries have been promoting different agendas and posturing for advantage. Because they believe that government leaders have shown a real lack of leadership towards advancing sustainable development, they conclude that leadership must come from the private sector. Therefore, the authors demonstrate how corporate enlightened self-interest to save dollars can simultaneously bring vast environmental improvement through pollution prevention and waste minimization.

Thomas Shen greatly expands this latter theme in his discussion of technologies for sustainable development. First he relates important criteria which define and distinguish sustainable technologies. Then he provides a comprehensive review of the status and prospects for sustainable technologies throughout several important economic sectors: energy, including fossil, nuclear and renewable sources; industry, including pollution prevention and clean technology; agriculture, including biotechnology; transportation and buildings. An important theme is how "designing for the environment" must incorporate product design and life cycle assessment in addition to efficient manufacturing processes and proper plant management practices.

Shen sees rather visionary roles for the engineering and design professions, government and the information and communications technologies. He believes that an environmental ethic must be widely adopted and that government must encourage sustainable technologies through proper regulation, use of incentives and partnerships with the private sector.

John Warren is also concerned with "operationalizing" sustainable development and incorporating principles of sustainable development within programs, policies and technology assessments. Using a systems approach, he defines a set of principles which characterize sustainable development and the effects of human actions over temporal and spatial scales. He then addresses the question of how we know whether we are progressing towards sustainability by analyzing sustainability indicators, including specific applications for global, national and local geographic scales. Finally, Warren discusses a comprehensive framework of sustainable development questions; these can be adapted by organizations which seek to incorporate sustainability into analysis, planning and decision making in accordance with the values of the organization.

Within the United States, some of the most critical immediate concerns for sustainable development are found within the state of California. Victoria Evans discusses how policies for sustainability must be developed. She then concludes this book with a case study of how government and regulatory bodies at different levels in California are attempting to cope with one critical problem of sustainability: air quality. This case study is particularly noteworthy and illustrative because it combines so many of the elements of sustainable development, including environmental limits, population, efficient use of resources, sustainable engineering technologies, education and social changes.

Evans notes that various engineering technological measures have been instituted to reduce automobile emissions, but that the dispersed settlement and land use patterns, population growth and vehicle growth have required new approaches also. Thus she describes recent efforts to influence land use and locational patterns and behavioral patterns to reduce the number of automobile trips and vehicle miles. The discussion of how different levels of government are attempting to work together within a legal/constitutional framework is noteworthy; the problems of data analysis are formidable, and both serve to indicate how much remains to be done to reach sustainability.

Despite the unique perspectives and experiences of the respective authors, several convergent themes emerge. Directing economic growth towards a sustainable future will require an interdisciplinary and systems mode of thinking. Energy, natural resources and materials must be used more efficiently and pollution must be prevented. Technologies and products must be designed according to well-defined criteria for sustainability. The private sector must become active in promoting an environmental ethic and must receive better regulatory and economic incentives. Changes in thinking and social behavior must accompany the development of an environmental ethic and sense of equity. Different levels of government will have important roles; ways must be found to incorporate land use planning at regional, urban and local levels to preserve landscapes, ecological diversity and ambient environmental quality. And we must develop useful indicators and data bases to monitor progress towards sustainability.

I would like to thank all of the co-authors for their enthusiasm and cooperation towards providing provocative manuscripts for this project. Finally, the vital support and understanding of my wife, Darlene, throughout has been much appreciated.

F. Douglas Muschett

1

AN INTEGRATED APPROACH TO SUSTAINABLE DEVELOPMENT

F. Douglas Muschett

As a paradigm and important environmental theme, "sustainable development" is puzzling. On the one hand, the term means what it says; sustainable development means economic development and a standard of living which do not impair the future ability of the environment to provide sustenance and life support for the population. On the other hand, it is more difficult to envision all of the forms and implications of "sustainable development" to relate one's professional career or personal lifestyle to its pursuit.

F. Douglas Muschett is principal of F. Douglas Muschett and Associates in Rochester, New York. He has held technical positions with environmental consulting firms and at Resources for the Future, Washington, D.C., and served as a professor of environmental policy and resource management at Syracuse University. His major interest has been resolving multiple objectives of economic development and environment through interdisciplinary analytical and policy approaches, about which he has authored numerous articles. He holds a B.S. in geography from the University of Michigan, an M.S. in meteorology and air pollution from Penn State University, and a Ph.D. in environmental systems and economic geography from the University of Michigan.

Perhaps part of the difficulty comes from the fact that "sustainable development," and the world in which we seek to practice it, cuts across and integrates many diverse disciplines. As humans inhabit and use the natural environment to improve a standard of living, they utilize a large variety of technologies and act, within the constraints of their culture, to transform the environment around them. However, in the twentieth century age of what I call "microspecialization," it is often difficult to see the whole system and how the parts are related to the whole. Therefore, much of what follows in this chapter underscores the importance of an interdisciplinary, "systems" approach in order to treat both fundamental problems and special situations.

In a very real sense, my goal of trying to relate what constitutes sustainable development is very humbling. We recognize that the objectives of sustainable development are to provide for the economic well-being of present and future generations and to maintain a healthy environment and life support system. However, no one truly knows what sustainable development is because we really cannot point to any examples where it has occurred. The wealthier industrial countries do not know about the "sustainable" part and most of the rest of the world does not know about the "development" part. Unfortunately, as we note below with some historical examples of the decline of civilizations, it is easier to cite where it has *not* occurred.

Moreover, it is impossible to lose sight of the fact that sustainable development is not strictly a problem of science or engineering or economics or proper management. The roots are found in values, ethics and culture of both developed countries and developing countries.

This chapter strives to articulate a better, integrated understanding of the imperative for and the many elements of sustainable development. At the same time, although I have attempted to be suggestive about some of the changes and approaches which will be needed, it is not an "action plan" for how to achieve sustainable development.

HISTORICAL PERSPECTIVES ON SUSTAINABLE DEVELOPMENT

Undoubtedly, prior to the highly publicized June 1992 United Nations Conference on Environment and Development (UNCED) in Rio de Janeiro, relatively few people had heard of the term "sustainable development." Since that time, although it is not exactly a household word, there has been rapidly growing interest among international organizations, the research community, environmental groups and professionals, and business to learn about "sustainable development," to promote it and, in some cases, to get in on the "next wave" of environmental concern.

Lessons from Other Civilizations and Societies

Although the term may be new, sustainable development is not a new phenomenon or concern. On the contrary, the impetus for our present concern dates back thousands of years, as so well illustrated by Dale and Carter in their compelling book, *Topsoil and Civilization*.[1] Two rather dramatic and insightful examples are the civilizations of North Africa, in the vicinity of ancient Carthage (now Tunisia) and Egypt, barely one thousand miles to the east.

At the height of its civilization and power, Carthage had over one million inhabitants and had an abundant food supply from the cultivation and grazing in the fertile lowlands between the coast and Atlas Mountains. Once Rome conquered Carthage and decided to make Carthage a colonial food supplier for the Roman Empire, a cycle of irreversible land degradation began, which impoverished people through history to the present. Rome opted for intensive cultivation with maximum yield per acre and when the fertility began to decline planted even more intensively to "make up" the declining yield. As productivity naturally declined even more, Rome spread cultivation and grazing into marginal and upland areas, triggering a cycle of erosion and declining productivity which ultimately ruined the land forever.

In contrast, civilization in Egypt persisted from the time of Cleopatra until the twentieth century on a "sustainable" basis; the annual spring flooding of the Nile provided both water and a replenishment of soil nutrients. Ironically, now, in the twentieth century, with the construction of the Aswan Dam, this stable system is in decline. In addition to a decline in soil fertility, which had to be supplemented by artificial soil fertilizers, there have also been many other well-documented, severe impacts upon health, sustenance and ecology from the altered hydrology and saltwater intrusion into the delta region.

Similar examples abound on virtually every continent, from the time of ancient civilizations through the Middle Ages and Renaissance periods and to the time of the Industrial Revolution. European countries, ranging from Ireland to Switzerland and Spain, among others, suffered ravages of deforestation, overgrazing and resulting flooding and loss of fertility. Watt presents an interesting theory on the decline of Spain as a naval and world power due to the inability of its limited forest resources to sustain the demands for shipbuilding.[2] Moreover, powerful landlords ("meseta") ruined a vast portion of the central and southern plain through the massive annual "sheepwalks," which denuded the land, changed the soil structure and damaged soil fertility.

Recent Roots of Sustainable Development

In the United States, we have only to think of whaling, the buffalo and the Dust Bowl as historical examples of "*non*sustainable development." It is not widely recognized that the seeds of our present concern with sustainable development were first sowed around the beginning of the twentieth century during the first wave of environmental concern in the United States, as described by Stewart Udall in his classic book, *The Quiet Crisis*.[3] The nation's first forester, Gifford Pinchot, promoted "conservation" as a field of inquiry to determine how the national forests could best serve the nation's many competing economic interests without depleting the forests over the longer term. At the time, he was vehemently opposed by John Muir, a "preservationist," who, in response to widespread destruction of natural resources during the settlement of the nation, fought to establish forests and wilderness as refuges to preserve the physical stock of nature and the spirit of humans.

As part of the wave of environmental concern in the United States following Earth Day in 1970, air quality became a primary concern and air quality policy began to address "sustainable development"—although, of course, that term had not yet been used—through questions of how to balance air quality and economic development. There were at least three contexts. One was the (continuing) question of how to enable continuing economic growth and development in areas which do not meet ambient air quality standards. A second concern was to ensure that continuing growth and development do not cause unsatisfactory air quality at a future time (air quality maintenance). A third, still important, context was the "prevention of significant deterioration" in wilderness regions which had pristine air. Generally, these approaches prescribed "technological retrofits" to specific polluters by rationing small increments of clean air at a time.[4]

More fundamental and controversial questions about the roles of population, resource consumption, environmental pollution and technology surfaced in the early 1970s during the so-called "Limits to Growth" debate. Under sponsorship of the prestigious Club of Rome, research by a group of scholars projected dire future global environmental consequences from some simplifying assumptions and extrapolations about population and resource growth rates.[5] Much furor and controversy resulted when these *projections* became widely interpreted in the media as *predictions*. Because, too, these projections neglected the capacity of humans and technology to adapt—about the same time as the "Green Revolution" demonstrated a capacity to greatly increase food production—the work became discredited for a while.

At the same time, it is important to note that on the twentieth anniversary of their original study, the authors updated the results in a new book, *Beyond the Limits*.[6] Using recent data and trends, the authors reached the same conclusions but underscore that environmental decay and economic decline are not inevitable provided that growth in population and material consumption is not perpetuated and provided that there is a drastic increase in the efficiency of use of materials and energy through technological improvements.

Aside from these projections of the future, contemporary issues and experience—ranging from tropical rain forests and global climate change to the Gulf War to the rapid economic and population growth in some developing nations—point out the necessity to live within the carrying capacity of the earth's ecosphere, to make the global economies more efficient in the use of natural resources and to reduce population pressures. There are, in fact, "limits to growth," and it is vital to ask (1) what kind of growth is desirable, (2) what kind is not and (3) how to develop economic policy and environmental policy accordingly while maintaining consumer choices and a sense of equity within a market economy.

DEFINING AND UNDERSTANDING SUSTAINABLE DEVELOPMENT

At the 1992 U.N. Conference on Environment and Development in Rio, UNCED Principle #3 characterized sustainable development as "the right to development must be fulfilled so as to equitably meet developmental and environmental needs of present and future generations." UNCED Principle #4 further states: "in order to achieve sustainable development, environmental protection shall constitute an integral part of the development process and cannot be considered in isolation from it." These two principles, stated as part of the U.N. Conference Agenda 21, have some very profound implications for use and stewardship of natural resources, ecology and environment, as I discuss later in considerable detail.[7]

For present purposes, it is important to ask: What does it mean to "equitably meet developmental and environmental needs of present and future generations"? I suspect that international dissension within the United Nations over Agenda 21 indicates that the answer is far from complete. Nonetheless, the "spirit" of Principle #3 would seem to indicate a "fairness" in meeting the needs of all peoples in the present generation, a "fairness" in meeting the needs of future generations as well as the present generation and a "balance" between development and environmental preservation.

There is a tendency in official gatherings and communiqués and agency programs to focus upon areas of consensus and very specific "missions." Issues of controversy are swept aside, and the operating principle is that continued economic growth and new technology will solve problems of poverty and environment for all peoples. Notions of "social change" or "zero-sum" economics" and "sacrifice" are politically incorrect. Yet, as noted in the following section, there is a considerable body of scholarly thought and research which indicates that sustainable development must include a major transformation of society. Consequently, in listing elements of sustainable development (Table 1.1), I have included some of the more fundamental, root causes, as well as economic, environmental and technology dimensions which are more frequently mentioned.

Indeed, we cannot attain sustainable development without better technologies which will enable us to "stretch out" scarcer nonrenewable resources and to utilize renewable resources. Nonetheless, although the focus of this chapter and book is not on social change, it is important for environmental professionals, economists and ordinary citizens alike to recognize that there are limits to what can be accomplished by technology.

TABLE 1.1   Elements of Sustainable Development

- Population stabilization
- New technologies/technology transfer
- Efficient use of natural resources
- Waste reduction and pollution prevention
- "Win–win" situations
- Integrated environmental systems management
- Determining environmental limits
- Refining market economy
- Education
- Perception and attitude changes (paradigm shift)
- Social and cultural changes

Ethics and Culture

It would be impossible to try to define sustainable development without discussing the importance of ethics and culture. The subjects of ethics and culture tend to make many individuals, including scientists, engineers and politicians, very uncomfortable. After all, isn't every person entitled to his or her "pursuit of happiness"? What "right" do any of us have to tell another person how to live (unless, of course, that person happens to be a relative or close friend, in which case it is our inalienable right!)? And, anyway, isn't there a certain inevitability to progress and new technology, and isn't the "free market" the best judge?

These questions have been debated, of course, with respect to every social issue imaginable, but the point here is to emphasize that ethics and culture are no less important with respect to sustainable development than with respect to other issues such as birth control, gun control, redistribution of wealth, etc. Furthermore, upon a minimal amount of reflection, it becomes obvious that many other social issues are closely linked to "sustainable development."

The ethical dimensions of sustainable development are two-fold: (1) our relationship to fellow inhabitants of our country and planet and (2) our relationship to the land and plant and animal inhabitants of the world. If many environmental professionals are shy, there is no shortage of ethicists, theologians and environmentalists willing to address these questions.

Is it immoral that the United States has to import over one-half of its energy supply? Or that a child born into the culture of the United States will consume 30 to 40 times per capita the energy and natural resources of the "average" of the rest of the world and 200 times as much as several undeveloped countries? Anglican Archbishop John Taylor believes so. In his provocative book *Enough Is Enough,* Taylor discusses environmental theology; based upon Judeo-Christian theology, he offers practical guidelines for a more responsible consumerism which promotes personal fulfillment and sharing but also reduces unfulfilling, unnecessary consumption.[8]

When so much of the fossil fuels and critical mineral resources have to be imported by the United States and other Western countries, there is much that has to be reformed within our cultures to set us on a path towards sustainable development. Notwithstanding the importance and role of technology, economics and better management strategies, many scholars are convinced that the only real hope for sustainable development is a radical shift in society.

Calculations by Adler-Karlsson demonstrated that a doubling of the population of the poor countries increases the consumption of world resources by one-sixth as much as doubling of the population in rich countries.[9] Carol and John Steinhart have observed that "the best energy technology can do is make things tidier while we struggle to change our habits." Stivers argued for "a new world view involving a radical change of attitudes and values."[10] Birch and Rasmussen note that "history's testimony is that the most far-reaching change comes only with the combination of strong pressures, from within and without, and a compelling alternative vision."[11]

These statements were written during the 1970s. Perhaps the reader can ponder whether the United States is beginning to see strong external pressures in the form of situations involving the Gulf War, Haiti, Cuba, Mexico and Somalia and strong internal pressures stemming from structural economic changes and social alienation and decay. No one knows what the "compelling vision" will be, but many scholars have suggested that it will have to be something of religious proportions. Both Taylor and Birch and Rasmussen have suggested the (Old Testament) concept of "shalom," or wholeness and harmony in relationships with neighbor and creation.

From the perspective of developing countries, the essence of sustainable development is to promote development which (1) reduces the disparities in lifestyles and global consumption and (2) improves and maintains a healthful local environment and (3) then, *and only then,* contributes towards solving critical global environmental management of the global "commons"—such as global climate change, oceans and fisheries, and forests.

There tends to be a wide spectrum of environmental ethics ranging from a belief that all plants and animals are on earth to serve humans to a belief that all life is part of creation and must be respected and protected. These two polar views, held respectively by "conservationist" forester Gifford Pinchot and "preservationist" John Muir, were the source of much acrimony during the spread of the first environmental movement within the United States towards the end of the nineteenth century. It should be pointed out, however, that mainstream religious denominations and theologians generally proclaim an intermediate view that plant and animal life and natural resources are on earth to serve humans, but that we also have a "stewardship" responsibility to care for the earth and its life.

Many economists take the "utilitarian" point of view that other species do not have an intrinsic worth and that, therefore, ecological protection should be based upon whether the species or habitat provides a direct economic benefit or indirect benefit through maintaining an ecological system.[12] In reality, human civilization and its diverse cultures, from traditional hunters and gatherers to sedentary agriculture to manufacturing to high technology, have already caused the extinction of many species and are encroaching upon the habitat and threatening the survival of thousands of others. To help resolve future conflicts between land use and economic activities and the survival of habitat and species, I believe that it is important to further develop and implement a set of criteria for setting priorities in the protection of plant and animal species and habitat.[13]

The Interaction of Whole Economic and Natural Systems

Note, too, in Principle #4 that making environmental protection an "integral part of the development process" is much *different from* the traditional pattern of making economic decisions and then correcting the environmental impacts which may result. It is a critical aspect of sustainable development that the *interaction* and *feedback* between the economic system and the environmental system be evaluated so that development can proceed in ways which will *prevent* and *reduce* environmental impacts.

Let us illustrate by examining the conceptual model in Figure 1.1. The interaction of the natural system and economic system and the flows of materials and energy are illustrated. It is important to note that as the term is being used here, the "natural system" includes the ambient physical environment, ecosystems and natural resources. The "economic system" refers to the fac-

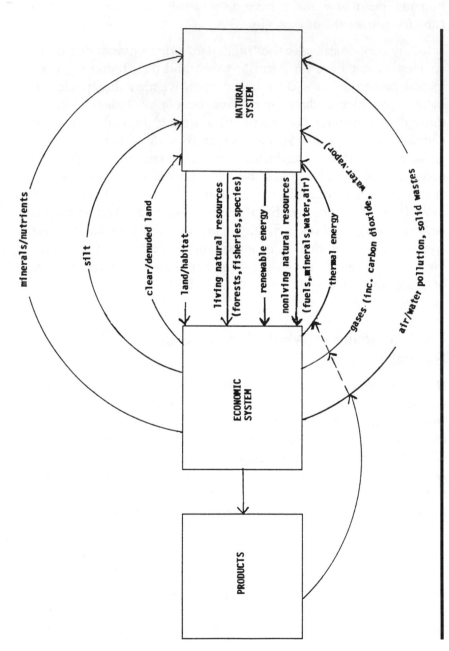

FIGURE 1.1   Conceptual interaction of economic and natural system.

tors of production for goods and services.[14] For purposes of conceptualization and discussion, the systems are generalized, but, as will be noted later, these "boxes" can be applied to specific economic sectors and products.

The purpose of Figure 1.1 is to illustrate the interactions in terms of (1) the kinds of input demands and stresses that the economic system places upon the natural system and (2) the waste outputs and stresses which the economic system places upon the natural system. With respect to the input demands posed by agricultural, industrial, commercial and residential economic sectors comprising the whole economic system, there are a few broad categories of stresses including: (1) the conversion of land and habitat to other uses; (2) ecological depletion and possible extinction of living species through harvesting, hunting, fishing and habitat conversion and (3) consumption of nonliving mineral and fossil fuel resources.

In terms of the outputs and stresses which the economic system places upon the natural system, again some broad categories are noted, including (1) air and water pollutants and solid wastes; (2) greenhouse gases, such as carbon dioxide, water vapor and other "trace" gases, and thermal energy; (3) "altered" land, which may have been cleared, denuded or paved and (4) silt, minerals and nutrients, resulting from erosion, runoff and decay products from both organisms and solid wastes.

From the standpoint of sustainable development, it should be observed that the importance of these impacts upon the natural system varies greatly geographically, dependent upon the existing states of both the natural environment and the economy. The United States has been slow to come to grips with its high per capita demands for natural resources. Although the high per capita consumption of energy and mineral resources is well documented, little attention has been given to the continuing loss of prime farmland. According to the U.S. Department of Agriculture, the United States lost 4 million acres of prime farmland to development during the period 1982–92, an increase of 18% in developed land compared to a 9% increase in population.

The loss of prime farmland, together with widespread soil erosion and soil profile changes, are examples of what can be considered an *"environmental deficit."* That is, analogous to an economic budget deficit which is repaid by and at the expense of future generations, environmental systems and natural resources are frequently consumed at the expense of future generations. In some cases, such as in the change of soil structure, the damage may be permanent. Similarly, with respect to soil erosion and certain nuclear wastes, damages cannot be reversed over many thousands of years. In other cases, such as toxic pollution of lake sediment and aquifers, the time frame may be almost as bad. On the other hand, in some instances, such as with certain forest regeneration and wetland regeneration, the restoration can be more timely.

Perhaps the first constructive step towards dealing with this "deficit" problem is the recognition that it exists. Towards this end, the U.S. Department of Commerce has recently developed a new indicator of Gross Domestic Product (GDP), which subtracts the consumption of natural resources and the costs of pollution and adds the benefits from improvements in environmental quality.

Fixing the Economic System Relations

There are several basic, underlying reforms needed throughout the whole economic system to reduce both the natural system inputs and the pollutant and waste outputs. First, with respect to the manufacturing sector, there must be more efficient product design and more efficient manufacturing processes and quality control (more on this later). There is a tendency to associate such modifications solely with the manufacturing sector. However, other reforms are also needed within the agricultural and commercial sectors as well. Careful farming methods, land use planning and proper construction practices can all reduce soil erosion. Better land use and transportation planning are necessary to reduce pollutant emissions and impacts. And new communications technology will likely reduce the amount of business travel, commuting and transportation pollution.

Another basic reform required to better integrate protection of the natural system within the macroeconomy is a shift or economic substitution for the inputs. For example, economic substitutions include a shift from fossil energy inputs to energy efficiency and renewable energy, a shift from a virgin resource to a recycled input and a shift from prime agricultural land development towards other lands. Achieving sustainable development in developing countries may even require some surprising shifts, such as from one form of renewable energy (wood biomass) to another form of renewable energy, or even a shift from wood to a fossil fuel (coupled with efficient energy use).

Economic Incentives

"The market" focuses upon profit and tends to allocate and reward investments with short-term paybacks. Some desirable policy outcomes, such as protection of the environment and conservation of natural resources, traditionally have not been achievable through reliance upon the market. Hence various programs of government regulation have evolved. Although government regulation has generally succeeded in meeting other, "noneconomic" goals, there has been increasing dissatisfaction from all sides. Slow response, slow adaptability to changing conditions, lack of innovation, excessive adjudication and expensive solutions—these have all been undesirable effects of government regulation.

The question becomes how to effect these kinds of more efficient product designs, manufacturing processes and economic substitutions of inputs. Invariably, as the world shifts more and more towards a global, free-market economy, these shifts and economic substitutions will have to be guided by economic incentives of various types. Closely related to the use of economic incentives, however, is manufacturer awareness of alternatives, consumer awareness and concern, changes in corporate culture and a concern for life cycle costs of processes and products. Many such "barriers" must be overcome before economic incentives can be fully effective.

Historically, within the field of environmental economics, economic incentives have been frequently viewed as either a form of payment or subsidy (incentive) or a form of tax (disincentive) or a combination of both, which is sometimes referred to as "push–pull." In a broader sense, however, more basic tools such as targeted investment tax credits can be important incentives to stimulate investment in more efficient manufacturing processes and to adopt the life cycle costing approach discussed below.

Other Western countries, more than the United States, have relied upon tax policy to reduce resource consumption and environmental impacts. Increasingly, though, environmental laws call for emission charges and fees as a disincentive for polluters. The present effort to revise tax policy to replace tobacco subsidies with tobacco taxes to pay for health care demonstrates a kind of "push–pull" tax policy; similar policies could eventually become more widespread as "sustainable development" issues such as global climate change, land use and prime farmland protection become more prominent.

However, it should be recognized that there are also various other powerful kinds of economic incentives beyond those defined by government economic policies. Undoubtedly, the strongest kind of incentive is a "bottom-line" cost saving through efficient operation, which means minimizing the input of energy and raw materials for operations and reducing waste. Related to this must be an increasing awareness and adoption of "life cycle costing" of products so that performance, durability and operation costs are taken into account for the lifetime of the product.

This neglect has generally hampered the adoption of state-of-the-art energy efficiency for homes and offices and factories. At least in Western countries, another important kind of economic incentive is the marketing and promotional value derived from consumer preference of "green" products.[15]

In the case of developing countries, there is *potentially* a particularly powerful economic incentive which could be used to promote sustainable economic development: international bank and lending policies. Unfortunately, this leverage is frequently not used in practice. Moreover, the increasing trend towards "privatization" with creative, private financing of projects in developing countries means that private profit interests can supersede public interests in sustainable development. For example, repeating the pattern of development in the United States, one of the more disturbing global trends in developing countries is the avalanche of electrical power plant construction, *without* strong accompanying energy efficiency programs, life cycle costing and full-cost accounting.

In fact, the preceding underscores the fact that there will always be a need for some regulation to promote public interests which would otherwise be overwhelmed by private interests. However, to the extent that incentives can be incorporated successfully within the economic system, environmentally sustainable development will be obtainable more quickly and at a generally lower cost than by regulation.

A NEW KIND OF ECONOMIC GROWTH

The U.N. World Commission on Environment and Development (WCED) observes, "Sustainable development requires a change in the content of growth, to make it less material and energy-intensive and more equitable in its impact."[16] In a global, "open" economy, the interactions between the economic and natural systems affect transfers at regional levels from one region to another. Thus, a related feature is that the economic demands from one region can cause problems of economic equity and human welfare and stresses upon the natural system in another region. For example, in developing countries, exports of cash crops and natural resources may reduce the land and natural resources available to sustain the local population and concentrate the wealth from exports among a relative few.

The above statement by the WCED raises issues which, in fact, are not new. Around the time of Earth Day 1970, both economists and environmentalists were discussing the question of changing the content of economic growth "to make it less material and energy intensive" and more equitable in its impact. In contrast to the traditional "cowboy" economy which fostered independence, recklessness and waste, economist Kenneth Boulding introduced the concept of a "spaceship economy." As the finite spaceship required the interdependency of the people and systems, a finite world requires people to work together within the limits set by the natural system and requires efficiency in our use of resources and care in our use of the environment.[17]

During a long career, economist E.F. Schumaker was concerned with economic development and equity—promoting the "right kind" of economic growth and factors of production which improve *local* employment and well-being. Although Schumaker was definitely out of the mainstream of a world which is concerned with maximizing growth rates, many of his tenets about culture, technology transfer and sustainable development projects are—finally, after his death—beginning to receive serious attention.[18]

Commoner's Simple Model

In a previous section, the interaction of the whole economic system and environmental system at a fundamental, highly aggregated level was discussed. However, let us suppose that the "economic system" box in Figure 1.1 now represents a specific product category such as electric power, automotive horsepower, plastics, wood pulp, etc. Then Figure 1.1 represents the stresses upon the natural system resulting from (1) the inputs demanded by the product category from the natural system and (2) the waste and pollutant outputs from the product category.

Commoner was concerned with the latter in an analysis of the origins of environmental impacts in the post-war U.S. economy.[19] For a large variety of economic goods (products), Commoner defined an "index of environmental impact" (which is really pollutant emissions) through the following relationships:

$$
\begin{aligned}
\text{Pollutants per Product} \quad = \quad &\text{Population} \quad &\text{(population)} \\[2mm]
&\times \frac{\text{Product Output}}{\text{Population}} \quad &\text{(affluence)} \quad &(1.1) \\[2mm]
&\times \frac{\text{Pollutants}}{\text{Product Output}} \quad &\text{(technology)}
\end{aligned}
$$

AN INTEGRATED APPROACH

Two examples of the kind of analysis performed by Commoner are given in Table 1.2. It is important to note in the above parentheses the interpretations of the three terms as given by Commoner. These interpretations are however, I believe, a bit simplistic. The second factor relating to per capita consumption of a good is indeed related to economic affluence, but consumer decisions are also related to culture and awareness. Similarly, the third factor which relates pollution to the amount of product produced is indeed related to technology and technological changes, but is also related to economics, regulation and corporate culture. Notwithstanding these criticisms, Commoner was

TABLE 1.2 Applications of Commoner's Model for Environmental Degradation

Synthetic Organic Pesticides: Environmental Impact Index

| | Index factors | | | Total impact index |
| | (a) | (b) | (c) | (a × b × c) |
	Population (1,000)	Crop production[a] / Population (crop production units/cap.)	Pesticide consumption / Crop production (1,000 lb/prod. unit)	Synthetic organic pesticides (million lb.)
1950	151,868	5.66×10^{-7}	3,326	286
1967	197,859	5.96×10^{-7}	8,898	1,050
1967 : 1950	1.30	1.05	2.68	3.67
Percentage increase, 1950–1967	30	5	168	267

[a] The crop output index is an indicator of agricultural production; 1957–59 average = 100.

Nitrogen Oxides from Passenger Vehicles: Environmental Impact Index

	Index factors			Total impact index (a × b × c)
	(a) Population (1,000)	(b) Vehicle-miles / Population	(c) Nitrogen oxides[a] / Vehicle-miles	Nitrogen oxides
1946	140,686	1,982	33.5	10.6
1967	197,849	3,962	86.4	77.5
1967 : 1946	1.41	2.00	2.58	7.3
Percentage increase, 1946–1967	41	100	158	630

[a] Dimension = NO_x (ppm) × gasoline consumption (gal. × 10^{-6}). Estimated from product of passenger vehicle gasoline consumption and ppm of NO_x emitted by engines of average compression ratio 5.9 (1946) and 9.5 (1967) under running conditions, at 15 in. manifold pressure: 1946, 500 ppm NO_x; 1967, 1,200 ppm NO_x.

Source: Originally published as Tables 3 and 7 on pages 46 and 57, respectively, in Reference 17.

able to present a rather convincing case, as in the examples in Table 1.2, for the relative importance in the changes in population, economic demand and technology as they affected the dramatic growth of pollution from different economic activities in the post-war period.

Aside from the specific sectoral analysis, part of the appeal of Commoner's work is that he attempted to analyze some fundamental causes for the dominant environmental problem of the time. It is perhaps interesting to also note that, in a far less technical fashion, an environmental theologian, Charles Birch, also developed a similar conceptual approach for analyzing root causes of environmental decay.[20]

Applying the Model to Sustainable Development

With respect to sustainable development, Commoner's simple approach can be adapted to provide some insights into how to "change the content of economic growth, to make it less material and energy-intensive." Although he was concerned with waste and pollutant outputs, a similar formulation can be used to examine *inputs* demanded by a specific product or economic sector from the natural system, as follows:

$$\frac{\text{Natural Resources}}{\text{Input per Product}} = \text{Population} \times \frac{\text{Product Output}}{\text{Population}} \times \frac{\text{Resource Input}}{\text{Product Output}} \quad (1.2)$$

Any resource of interest (e.g., energy, metals, wood, land, etc.) could be analyzed in this fashion to determine the relative importance of fundamental factors in the demand for inputs. For example, two such formulations could be:

$$\begin{array}{rl}
\text{Fossil Fuel Energy} \\
\text{Input to Steel} \end{array} = \text{Population}$$

$$\times \ \frac{\text{Steel Output}}{\text{Population}} \qquad (1.3)$$

$$\times \ \frac{\text{Energy Input}}{\text{Steel Output}}$$

or

$$\begin{array}{rl}
\text{Land Input} \\
\text{to Housing} \end{array} = \text{Population}$$

$$\times \ \frac{\text{Housing Output}}{\text{Population}} \qquad (1.4)$$

$$\times \ \frac{\text{Land Input}}{\text{Housing Output}}$$

Viewed in this manner, from the general formulation (Equation 1.2) above, any measure which serves to reduce the factors on the right-hand side will reduce proportionately the natural resource inputs required for a given economic product. The first observation is the direct importance of population in sustainable development. The coupling of the substantial populations in developing nations and their desire to become economic consumers like the Western nations is an emerging cause for global environmental concerns and a major driving force for sustainable development.

Therefore, it is essential for the Western countries, for reasons of both "fairness" and their own self-interest, to become "better mentors" in their consumption of natural resources. We can begin by using the above scheme to ask the following questions:

(1) Is it desirable or feasible to effect behavioral changes to reduce the per capita consumption of the product (second term)?

(2) What technical means are available to reduce the resource input per unit of product output (third term)?

Per Capita Consumption

Occasionally, in specific instances which are regarded as being important to the general health and well-being, there are efforts by the government or public interest groups to intervene to change consumer habits (e.g., smoking, energy conservation education, safe driving, eating habits or products from endangered species). In the short term, there are already many individual examples whereby an informed, aware consumer may wish to shift consumption from one product to another for reasons of health, economics or consumer satisfaction—and at the same time promote "sustainability"—a "win–win" situation.

For example, by reducing electric power consumption, the consumer saves money and reduces fossil fuel inputs. By reducing consumption of corn-fed beef for health reasons, a reduction in energy and agricultural chemicals is also brought about. By purchasing a smaller house, the consumer can save money, reduce maintenance and increase leisure time, and reduce natural resource consumption.

In a free society, any large shifts in consumerism, such as becoming a less materialistic and consumptive society, are necessarily dependent upon values and major social and cultural changes over long periods of time. Such changes must overcome a lot of skepticism among both economists and consumers about the nature of economic growth and what would happen to the consumer-based economy if there were a decided longer term shift towards a less materialistic society. For a long time, environmental quality—clean air and clean water, for example—was thought to be a "drag" on the economy and jobs. It was just not perceived that consumers could choose to demand more clean air and more clean water in the sense that they could demand other economic goods. Now, however, it is widely recognized that the demand for environmental protection creates jobs.

In the same way, in the long term, possible future shifts in consumer demand towards fewer but more durable goods and more services (e.g., "online," arts, recreation, etc.) are compatible with a healthy economy. This is not to imply that there is no concern about dislocations of products and jobs; for example, the U.S. automobile industry requires time for planning and adaptation. However, there is no reason to fear cultural changes and related changes in the economy over time. Historically, in fact, the interaction of culture, technology and the market economy has demonstrated that every product and service in the marketplace has its so-called "S" curve, featuring stages of rapid growth, slower growth, stability and decline.

Reducing the Inputs

It is important to note that the extent to which we are able to reduce the ratio of "material inputs to product outputs" over time, either by efficiency or substitution methods discussed below, is an important indicator of technological progress towards sustainability. Thus, in conjunction with the kind of product life cycle analysis discussed earlier, it is important to evaluate trends in this ratio.

More Efficient Use of Inputs

During the past decade, "bottom-line" priorities and competitive pressures have shaped an emphasis upon more efficient manufacturing processes which (1) use inputs more efficiently and produce less waste per unit of output and (2) have better quality control and produce less waste. More recently, the design of the product itself, size and packaging are becoming recognized as important means of reducing resource inputs. An emerging field of product "life cycle design" is studying ways to promote sustainability, including the use of component parts which can be recycled.[21]

Substitution of Inputs

Somewhat related to the previous concept of product design is another method of reducing resource inputs: a substitution of inputs. The idea is to substitute a more plentiful resource for a critical or less plentiful resource. There are several kinds of substitutions which are possible for manufacturing. Substitutions can include one nonrenewable resource for another (steel instead of tin), a renewable resource for a nonrenewable resource (wood or biomass-derived chemicals instead of petrochemicals), one renewable resource (maple wood) for another renewable resource (mahogany) or, as in the production of music keyboards, a renewable resource (wood) or even a nonrenewable resource (plastic) for an endangered resource (ivory).

Such substitutions are also critical for the economies in developing countries. In most cases, countries must utilize abundant local natural resources such as sand, stone, wood and fossil fuels for housing and fuel, respectively. However in other cases, they must try to protect diminishing natural resources, such as forests, from population pressures.

Despite the above examples as to how a substitution of inputs can promote sustainable development, the opposite is often true in practice. That is, for reasons of product economics, performance and consumer preference, there are frequently substitutions which utilize more critical resources and nonrenewable resources. For example, over the past two decades automobile construction has shifted away from steel towards aluminum and now plastic. Rubber tires are no longer made from natural rubber. Containers have shifted towards plastics.

This situation makes one particular kind of substitution increasingly important for sustainable development: a recycled input. As part of the emerging "life cycle design" noted above, products are being designed so that the components can use recycled materials and so that the components themselves can eventually be recycled. Economist Herman Daly suggested the ultimate, idealized goal of a "stationary state" economy which minimizes what he referred to as the "throughput" by reusing and recycling.[22]

A few products in which the United States is the world leader but which would never come to mind are iron and steel scrap metals and waste paper. These waste products have economic value in large measure due to the energy saved in using the waste scraps versus processing raw materials. So, the United States exports the waste products, and many countries, particularly the Asian countries, in turn produce finished products like steel, paper, autos and appliances. This is not to suggest that U.S. manufacturers are stupid or unaware; there are many complicated factors and domestic "barriers" to the use of recycled materials which led to this situation.

Increasing the Value-Added of the Resource

Computers and software, communications technology, aerospace, agricultural products, perhaps even environmental control technology—these are products in which the United States is a world leader. In most cases, these products share the fortunate economic trait of having a high "value-added." That is, above and beyond the economic value of the natural resources in the product—the cost of the silicon and metal comprising computer components is rather minimal—there is considerable value added to the product by the sophisticated technology, professional engineering and design, and worker skills.

In the context of sustainable development, one way of reducing the natural resource input per unit of product is to improve the quality and durability of the product; that is, to manufacture a "better," more expensive product. A second way, as we have noted above, is to develop technology and economic incentives which will facilitate greater use of scrap materials. A third way to increase the economic output from the resource is to shift the product itself to another product. For example, if there are limitations upon the amount of forest that can be sustainably harvested, instead of exporting timber or lumber (the next higher valued product above timber), it may be desirable to produce furniture or doors or cabinets or prefabricated housing.

Pollution Prevention and Waste Minimization

Before concluding this section, recall that I began with an observation from the U.N. World Commission on Environment and Development that sustainable development requires a change in the content of growth, to make it less material and energy intensive and more equitable in its impact. Therefore, I discussed some approaches towards reducing the inputs of natural resources required for individual economic products. However, as we have noted previously, actions taken to reduce the economic inputs will also reduce the wastes and pollutants produced by the economic system.

Thus there is a close relation between the above approaches and the concepts of "pollution prevention" and "waste minimization" discussed in other chapters. Pollution prevention emerged during the 1980s as perhaps the most important environmental paradigm of the decade; it was originally developed by a handful of corporate giants acting in their enlightened self-interest to save money on the costs of production and the costs of air and water pollution control.[23] Methods of pollution prevention include changing industrial processes, changing the inputs to industrial processes, reusing industrial wastes, using industrial energy more efficiently and changing product design. Subsequently, the concept expanded to embrace waste minimization for analogous methods of reducing solid and hazardous wastes.

One specific related issue deserves mention, however. Since considerable pioneering research undertaken during the 1970s at

Resources for the Future in Washington, D.C., it has been recognized that beneficial, economically efficient trade-offs among the components of the environmental media can be obtained with certain changes in manufacturing processes. For example, a decrease of 50% in water organic emissions might be achieved with a process that increases air sulfur dioxide emissions by 10%.

Because the evolution of regulatory policy in the United States has been towards singular pursuit of specific air and water pollutants, an efficient, integrated "multimedia" presents formidable legal obstacles. Despite some recent attempts towards "multimedia" management concerns, there is still a long way to go. Hopefully, other countries can develop more flexible approaches in their emerging environmental management programs.

Win–Win Strategies in Sustainable Development

In the course of other discussions I have alluded to the fact that there exist possible "win–win" situations; that is, taking actions which will meet more than one objective at the same time. I believe that an important component of sustainable development in the near term should and will be efforts to identify and promote "win–win" development strategies and policies which can simultaneously help meet both development objectives and environmental objectives. Certainly, these "win–win" types of policies are more capable of generating public, private and political support than policies which are perceived to simply restrain development. And some of these policies can also be helpful immediately while we grapple with changes in our technology, economic system, thinking and behavior.

There are a few specific categories of "win–win" situations which deserve more explicit mention than previously given. One category is agriculture, a form of renewable resource. In some developing countries, land is used almost exclusively for the production of so-called "cash crops." Another possibility, which promotes both renewable resources and economic development, as Brazil has shown, is to produce biomass-derived chemicals and fuels.

However, there are also important benefits from shifting some land use from cash crops, generally owned by large landowners, to more food production with individual land ownership. The population will be better fed, and better land stewardship is generally possible. Of course, "cash crops" are an important economic base which serves to pay for imports and taxes; however, the wealth tends to be concentrated and much of it flows for purchase abroad rather than stimulating the local economy. Often it will also be necessary to effect a degree of land reform and transfer of power and wealth.

From a U.S. perspective, our present agricultural system can be characterized as an amazing success story in agricultural output, but one which is completely nonsustainable with present methods. The vast increases in agricultural output and consumer choice have come at the expense of methods which have cost approximately one-half of the topsoil and have required a "subsidy" of about ten calories of fossil-fuel input for farming, processing, distribution and preparation for each calorie of food output. In the long run, it is also desirable that agricultural policy help effect a return of the farming occupation and the "family farm." This will help to provide economic opportunities and improve rural economies and also likely will promote better care of the land (a family tends to be interested in maintaining and not "depreciating" the land).

A second kind of "win–win" situation for sustainable development, an extension of a prior discussion, is to manufacture new products from recovered waste products—solid wastes (rubber, plastics, scrap metals, papers), agricultural and organic wastes (in excess of those needed to maintain soil fertility) and animal and human wastes. Many states, such as New York, now have considerable departments within their respective economic development agencies which are dedicated towards promoting and providing incentives towards these goals and overcoming some of the traditional market barriers. Many successes are *beginning* to occur, such as the recovery and remanufacture of plastic products like packaging materials.

Energy Efficiency and Renewable Energy: The Ultimate Win–Win Situation

Clearly, energy is a critical factor in sustainable development. On the one hand, the high per capita consumption of fossil fuels by the United States and other Western nations is nonsustainable, and on the other, the developing countries are seeking to become more like the West in lifestyle and technology. Energy costs are an important component of industry competitiveness and consumer expenditures, and the environmental impacts of fossil-fuel energy are far-reaching.

With respect to renewable energy, although many technologies of the future, such as solar photovoltaic power, are not generally economically competitive now, there are a variety of niche applications which are. Wind power is particularly well suited to powering small communities and irrigation, and solar photovoltaics is well suited for remote areas away from a transmission grid. Some forms of biomass wastes, ranging from agricultural to manures, are also a renewable form of energy, undoubtedly more widely used in the developing countries than the developed countries. The "piggybacking" of certain alternate sources of energy is also possible.[24]

Several years ago at a conference, I made the comment that energy conservation is an important strategy to combat air pollution; in response, I well remember receiving a number of blank stares of bewilderment. Thus, despite the fact that the major source of conventional air pollution, photochemical smog and acid precipitation alike is fuel combustion, these issues have been widely perceived as pollution issues, not as energy issues. Consequently, environmental management strategies have successively focused upon these issues as separate problems requiring separate programs and technologies to clean up pollution from fuel combustion. However, simultaneous "win–win" situations exist by preventing pollution through more efficient use of energy.

Fortunately, perception of the emerging greenhouse issue of global warming is different, although its major cause, energy consumption, remains the same. This situation provides an important environmental "win–win" opportunity; energy efficiency measures taken to reduce the consumption of fossil fuels will not only reduce the emission of carbon dioxide, a major greenhouse gas, but will simultaneously reduce the emissions related to several other atmospheric pollution problems as well.

Energy efficiency has several economic benefits as well. Because energy efficiency reduces the costs of energy, the local and national economies are helped in several ways: (1) domestic manufacturers can be more competitive, (2) corporate and consumer disposable income is increased, (3) more of the latter is spent in the domestic economy, (4) the vulnerability of the domestic economy to international oil prices is reduced and (5) new jobs are created in energy efficiency. New technology for lighting, industrial motors and household appliances allows truly amazing energy savings of 30 to 60% now.[25]

Because the other resource conservation, environmental and economic reasons for energy efficiency are compelling in their own right, it is not necessary to wait until all the facts are in to begin to take action against global warming. Moreover, a widely touted U.S. Environmental Protection Agency report concluded that delaying by a few decades could increase the global warming commitment by 30%.[26] In response to the 1992 United Nations Conference on Environment and Development in Rio, the United States is taking a leadership position by adopting a voluntary program, the President's Climate Change Action Plan.[27]

Although this program is relatively modest, it is a beginning. A compelling goal for sustainable development and global climate policy alike will be to further integrate environmental, energy and economic policy to provide "win–win" situations. As I have written elsewhere, there are three components which are essential towards this goal:[28]

(1) Adopting a holistic environmental management framework for related environmental problems and solutions

(2) Fostering a creative combination of regulation, incentives and penalties to guide consumer, industry and the marketplace

(3) Research and development initiatives that emphasize the *utilization,* as well as the development, of energy efficiency and renewable energy technology

SUSTAINABLE DEVELOPMENT AND THE AMBIENT ENVIRONMENT

The Role of Assimilative Capacity in Environmental Management

In the previous section, I discussed sustainable development from the perspective of trying to change the content of economic growth to become "less material and energy intensive and more equitable." As necessary as this objective is towards the goal of achieving sustainable development, it is not sufficient. If the broad objectives of reducing resource consumption inputs and reducing pollutant outputs are achieved, neither will be achieved absolutely. Nor will population growth be stabilized in the foreseeable future. Hence, there will always be stress upon the ecology and ambient physical environment.

Originally, the concept of a "sustainable yield" was applied to enable the harvest of ecological renewable resources, such as forests and fisheries, at the same rate as nature (assisted by human management) was able to replenish.[29] There are limits as to what nature will permit without damaging the ecological system and resource base. Similarly, environmental scientists have come to recognize that the physical, chemical and biological characteristics of the ambient environment determine the ability to accept, dilute, diffuse and transform pollutants; this "assimilative capacity" limits the amount of pollution tolerable without causing damages.

This principle holds whether one is considering a very localized thermal pollution plume in a river, a regional air pollution problem or a global climate change. In general, as the geographical scale increases, however, the complexities and interactions of natural scientific processes also increase. Thus, to achieve sustainable development, in all cases we must be concerned with the "assimilative capacity" of the environmental system, which in turn determines the "carrying capacity" for supporting population and economic activity and resulting pollutant emissions.

The general framework for environmental quality management, shown in Figure 1.2, has been well established in the United States and Western countries for air and water quality management. There is often an "iterative" process which examines different strategies, the resulting spatial patterns of emissions and the modeled ambient concentrations which would theoretically result as a result of the assimilative capacity. Ultimately, one or more environmental strategies are selected and implemented. This general framework is useful not only in the United States and Western countries, but also in developing regions around the world. However, as noted below, a more creative mix of environmental management strategies will be required than has generally been adopted in the United States.

Air Quality Management and Sustainable Development

Engineering control strategies have generally been highly effective in reducing air pollutant emissions, often in excess of 95 to

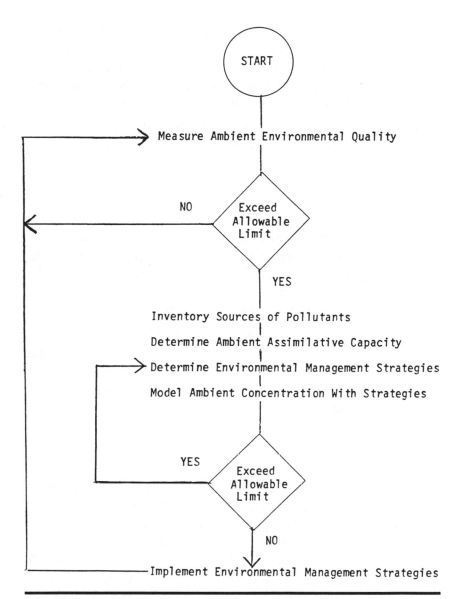

FIGURE 1.2   General framework for air and water quality management.

99%, and in meeting standards for air quality. Although engineering control strategies will continue to be widely used, many are more costly than pollution prevention and other strategies; many countries will face difficult choices among expenditures for social well-being and must pursue other strategies as well.

There are also some situations in which the combination of a high density of population and economic activity, together with a poor ambient air quality assimilative capacity, defy a solution based strictly upon engineering controls of pollutants. Over a decade ago, as part of a course I taught in environmental management, we were able to demonstrate using unsophisticated, "back of the envelope" calculations that there was not a conventional technological solution to the smog problem in Los Angeles. The population increase, coupled with more automobiles traveling more miles, simply overwhelmed the rate at which air quality improvements were made from the "turnover" of new vehicles with better technology replacing older vehicles.

Other important air quality management strategies include (1) land use planning strategies, (2) transportation planning and (3) energy management strategies. There are several types of land use planning strategies which, incidentally, *do not* mean that Big Brother is dictating exactly how a land owner is able to use the land.[30]

Infrastructure strategies, including the placement of natural gas pipelines, major highways, high-speed rail and sewage treatment facilities, are highly effective on both regional and local scales in influencing where economic growth and development take place. Locational strategies and incentives can be used to influence where industrial development or residential development takes place; these strategies are currently being used in locations such as Mexico City and Beijing to both reduce the density of pollutant emissions and reduce the exposure of the population. In smaller cities, green space and open-space strategies can be used at local geographic scales to improve the assimilative capacity for air pollutants.

Transportation strategies to reduce automobile emissions have been widely used within many major urban areas of the United States, with varying degrees of success. Strategies include carpooling, limiting highway access, mass transit, light rail, parking fees, road tolls, bicycle paths and flexible work scheduling. Many European countries have been more effective in reducing their dependence upon the automobile for transportation than has the United States (another example of the importance of culture in sustainable development). In Europe, however, the high population densities and congestion in cities, and relatively short distances for intercity travel, tend to reduce the attractiveness of automobile commuting and to enhance alternative modes of transit. (Some very provocative ideas on land use and transportation planning are presented in Chapter 6.)

Energy management strategies include "fuel-switching" to cleaner fuels, adoption of energy-efficiency strategies to reduce the demand for energy, cogeneration of electricity along with industrial process heat and the development of alternative and renewable sources of energy, including biomass and agricultural wastes, geothermal, wind and solar photovoltaic energy.[31] Oddly, whereas "fuel-switching" was widely recognized as a major pollution control strategy during the 1970s, energy efficiency and energy conservation still are not.

Water Quality Management and Sustainable Development

It must be emphasized that much, if not most, of the preceding discussion related to air quality also applies to water quality. That is, bodies of water—rivers, lakes, estuaries—have an (highly variable) assimilative capacity which is dependent upon the physical and biochemical processes which dilute, mix and transform pollutants. (Groundwater, such as aquifers, is a special case with very limited assimilative capacity because of limited water flows and turnover.)

Dependent upon the amount of pollutant discharge into the water and the assimilative capacity, the resulting concentrations of pollutants in the water and in the tissues of aquatic organisms will determine whether the body of water is fit or unfit for human consumption, aquatic life, commercial fishing, recreational purposes or industrial use.

Moreover, there are two notable constraints that the environmental system poses in relation to water quality. First, river flows tend to be extremely variable from season to season, and coupled with withdrawals for human use, there is a severe upper limit on assimilative capacity. Second, the increasing "bioaccumulation" of pollutants in successively higher levels of the food chain also severely limits the allowable concentration of pollutants in the water.

The combination of these limiting ambient conditions, together with a high density of population and economic activity, can make sustainable development very difficult to achieve. Many developing nations suffer widespread poor water quality due to a combination of high population density, widespread pollutant discharge from natural resource extraction and processing, increasing industrialization and inadequate assimilative capacity.

Since the age of industrialization began, most countries of continental Europe have experienced widespread poor water quality in surface rivers. Within the five river basins comprising the Ruhr district in Germany, a unique and fairly complex system of water quality planning and "stream specialization" had evolved by the 1960s. For example, the Ruhr River was maintained in a state of water treatment and ambient water quality suitable for water supply and recreation, whereas a parallel river, the Emscher, had been designated for untreated pollutant discharges and the dilution of pollutant wastes carried downstream.

Although the Ruhr district, one of the most concentrated industrial and population centers of the world, is hardly an example of sustainable development, this example does serve to underscore the importance of combining engineering and technology strategies and land use planning to achieve sustainable development. In some cases, it will also be possible to implement "win–win" situations; that is, infrastructure and locational strategies can be devised that will protect both air quality and water quality. Perhaps the case was best stated by landscape architect Ian McHarg in the title of his book *Design with Nature.*

Ecological and Life Support Issues

The preceding principles and discussions of this chapter are offered with the goal of nurturing some general, integrated approaches which can be used within the market economy and public policy to begin to achieve sustainable development as previously defined. Yet, there are special, critical problems which will require special, intricate responses, very possibly with much trial and error and disappointment along the way. Some of these critical problems include:

(1) Species and habitat protection, as previously discussed

(2) Global agriculture—emphasis upon feeding the population (not just raising cash crops) and agricultural methods which use indigenous and renewable resources and maintain soil structure and fertility

(3) Tropical rain forests—forest management for suitable economic uses, such as harvesting of fruits, nuts, hides, plants and lumber (subject to proper management of tree species, density and location)

(4) Global climate change—as previously discussed, identifying a continuum of possible environmental management responses to reduce emissions of greenhouse gases and implementing programs which make sense to do now for other reasons (e.g., "win–win" situations such as methane gas

recovery for fuel, composting and energy management)

These problems are well established and of sufficient complexity to be beyond the scope of a detailed analysis here. However, in the spirit of sustainable development and UNCED Principle #4 (discussed above), it is important to emphasize that the "integration of development and environmental protection" will again require an interdisciplinary approach. The close working cooperation of scientists, natural resource specialists, economists, geographers and planners will be needed to formulate, evaluate and implement development and protection strategies within a given culture which either do not diminish the life support systems or can provide reasonable trade-offs.

Holistic skills and perspectives are needed to "ask the right questions" to "obtain the right answers." Creativity is required for development strategies which meet the needs of the local population and utilize local resources without exceeding environmental limits.

CONCLUDING REMARKS

In this chapter, I have presented an overview of sustainable development and the relationship between the natural systems and the economic system. It is important to realize that "sustainability" from the standpoint of either the availability of natural resources to meet the needs of the world's population in an equitable manner or from the standpoint of environmental protection is really two sides of the same coin. That is, an integral part of the solution to both involves finding ways to limit per capita natural resource consumption in both developed and developing nations and ways to substitute renewable resources and "waste products."

As depicted in Table 1.1, sustainable development includes several elements. There is widespread agreement that population stabilization in developed and developing countries alike must be a first priority. Similarly, it is widely agreed that more sustainable technologies must be developed and employed.

However, it is difficult to say which of the other elements is more important than the others. Clearly though, some of the objectives will take relatively long periods of time to effect: for example, population stabilization, refining market economies, adopting a systems thinking perspective, integrating environmental management approaches, education and changes in social thinking and cultural behavior. Hence it is important for all nations to begin initiatives, including research and pilot programs, incentives and transfer of appropriate technologies, which will support and effect these changes. An important next step is to set specific priorities and to establish programs to meet global and national sustainable development concerns within the national and local context of needs.

This is not to suggest, however, that all actions must await the development and funding of grandiose master plans. "Grass-root" movements are important to effect perception and attitude changes, which will lead to social and cultural changes. Personal, individual reflection and response to a sustainable ethic are vital.

It is also important for the private sector—where capital, information and expertise are concentrated—to grasp the business opportunities presented by sustainable development. By taking reasonable risks to develop new tools of analysis, products and services, which promote better management of natural systems, efficiency and reuse, the private sector can stimulate important new markets which will aid the transition.

ACKNOWLEDGMENTS

I would like to express my gratitude to Mitch Baer, American Petroleum Institute, and Chuck Marshall, JACA Corporation, for their review and critique of this manuscript and helpful comments.

NOTES

1. V.G. Carter and T. Dale, *Topsoil and Civilization,* rev. ed., University of Oklahoma Press, Norman, 1974.

2. K. Watt, *Understanding the Environment,* Allyn and Bacon, Boston, 1982, pp. 352–353.

3. S. Udall, *The Quiet Crisis,* Holt, Rinehart and Winston, New York, 1963.

4. It should be noted that various air quality planning tools, including air quality modeling, emission density zoning and optimization modeling, were developed and are available for applications of sustainable development in which the carrying capacity of the airshed is threatened. For additional information, see J.R. Kurtzweg and C.J. Nelson, "Clean air and economic development: an urban initiative," *J. Air Pollut. Control Assoc.,* 30:1187–1193, 1980; F.D. Muschett, "Clean air with economic growth: optimization modeling," *Environ. Manage.,* 6:145–154, 1982.

5. D.H. Meadows et al., *The Limits to Growth,* Universe Books, New York, 1972.

6. D.H. Meadows et al., *Beyond the Limits,* Chelsea Green Publications, Post Mills, VT, 1992.

7. Agenda 21, UN Conference on Environment and Development, Rio de Janeiro, 1992.

8. J.V. Taylor, *Enough Is Enough,* Augsburg Publishing House, Minneapolis, 1977.

9. A.M. Thunberg, "The egoism of the rich," *Ecumenical Rev.,* 26(3):460, 1974.

10. R.L. Stivers, *The Sustainable Society: Ethics and Economic Growth,* The Westminister Press, Philadelphia, 1976.

11. B.C. Birch and L.L. Rasmussen, *The Predicament of the Prosperous,* The Westminister Press, Philadelphia, 1978, p. 73.

12. It should be observed, however, that such decisions run the risk of being based upon present knowledge. If species and genetic

pools are eliminated, then the possibilities for discovering future benefits to humankind are also.

13. For example, criteria might include the "endangerment" of the species, the "uniqueness" of the species, the importance of the species in the food chain and ecological system, the ecological productivity of the habitat, the scarcity of the habitat and other alternatives for human settlement and use.

14. Note, too, that the product outputs are placed in a separate box, apart from the economic system, because products are a form of "storage." The return of matter from both "durable" and "nondurable" goods occurs over time, whereas pollutant wastes are returned immediately to the natural system.

15. Although an environmental ethic in Western countries is becoming an important part of purchasing decisions, it is frequently difficult for consumers to make intelligent decisions. At one level of analysis, product labeling and information are often confusing and misleading. However, at a deeper level of analysis, there are many factors to consider when trying to make a "sustainable" consumer decision. For example, it is not obvious whether vinyl siding for a house or wood shingles with paint is a better choice or whether an electric power mower or gasoline power mower is more desirable.

16. United Nations World Commission on Environment and Development, *Our Common Future,* Oxford University Press, Oxford, 1987.

17. K. Boulding, "The economics of the coming spaceship earth," in *Environmental Quality in a Growing Economy,* H. Jarrett, Ed., Johns Hopkins Press, Baltimore, 1966.

18. E.F. Schumacher, *Small Is Beautiful,* Harper and Row, New York, 1973.

19. B. Commoner, "The environmental cost of economic growth," in *Energy, Economic Growth and the Environment,* S. Schurr, Ed., Johns Hopkins Press, Baltimore, 1972.

20. C. Birch, "Creation, technology and human survival: calls to replenish the earth," *Ecumenical Rev.,* 28(1):70, 1976.

21. G. Keoleian and D. Menerey, "Sustainable development by design: review of life cycle design and related approaches," *Air and Waste,* 44:645–667, 1994.

22. H. Daly, Ed., *Towards a Steady-State Economy,* W.H. Freeman, San Francisco, 1973.

23. F.D. Muschett and M. Enowitz, "The changing pollution control industry," *Pollut. Eng.,* 18:44–47, 1986.

24. For example, in producing alcohol fuels, the distillation process requires a large input of energy to reduce the water content of the fuel; solar energy could potentially replace fossil energy for this purpose.

25. A. Lovins, "Abating air pollution at negative cost via energy efficiency," *J. Air Pollut. Control Assoc.,* 39:1432–1435, 1989.

26. D. Lashof and D. Tirpak, Eds., Policy Options for Stabilizing Climate, U.S. Environmental Protection Agency, Washington, D.C., 1989.

27. B. Clinton and A. Gore, Jr., The Climate Change Action Plan, Executive Office of the President No. 93-0624-P, Washington, D.C., 1993.

28. F.D. Muschett, "Global warming calls for changes in public climate," *Forum Appl. Res. Public Policy,* 6:44–54, 1991.

29. In practice, this is a kind of "dynamic equilibrium" in the long run because natural factors such as climate and ecological productivity vary from year to year. It should also be noted that when the harvest rate exceeds the replenishment rate, a new state of equilibrium with a lower resource base is reached.

30. Unfortunately, during the Reagan administration there was a backlash against the notion of land use planning and the Land Use Planning Branch within the U.S. Environmental Protection Agency was terminated. During the 1980s the "deindustrialization" of the United States, together with the large number of plant closings, probably also meant that the immediate applications of land use planning for air quality were more related to transportation issues than industrial-related air quality.

31. From the standpoint of sustainable development, one has to be rather analytical and cautious about what is sometimes called "alternate" sources of energy. Certain "alternate energy" forms may be counterproductive, such as wood-fired power plants which compete for limited forest resources or so-called "resource recovery" plants which are designed to burn garbage without separation, reuse and recycling of nonorganic materials. Even some forms of solar collectors are very resource-intensive in terms of the amount of glass, metal and land consumed in relation to the amount of energy produced.

2

AN ECOLOGICAL PERSPECTIVE ON SUSTAINABLE DEVELOPMENT

C. Lee Campbell and Walter W. Heck

Sustainable development has been defined as "development that meets the needs and aspirations of the present without compromising the ability of future generations to meet their own needs."[1] In practice, sustainable development is a multifaceted concept, which can be viewed from many perspectives but must be defined in terms of the time period being considered (i.e., years, decades, centuries or millennia) and the proportion of ecosystem structure, function and composition that is to be maintained.[2] The perspective adopted to discuss sustainable development depends on one's personal viewpoint and discipline. Realistic viewpoints could be, for example, those of a resident of a developed or developing country, of an environmentalist or industrialist, or of a scientist or policymaker. Disciplines with a vested

Dr. Campbell is a collaborator with the USDA Agricultural Research Service, Air Quality Research Unit, 1509 Varsity Drive, Raleigh, NC 27606 and Technical Director of the Agricultural Lands Resource Group of the Environmental Monitoring and Assessment Program (EMAP); he is also Professor of Plant Pathology at North Carolina State University. Dr. Heck is a Plant Physiologist (retired) with the USDA Agricultural Research Service, Air Quality Research Unit and Associate Director of EMAP for Terrestrial Systems; he is also Professor of Botany at North Carolina State University.

interest in the concept of sustainability include agriculture, ecology, economics and sociology, among others. One might raise the question as to whether sustainable development and ecological sustainability are compatible concepts.

The ultimate question with regard to sustainable development asks whether we can sustain the development of the resources of Planet Earth in order to support a human population today without undermining the potential of the planet to support future generations. With a current human population of approximately 5.5 billion to be supported today, growing at an annual rate of 1.7%, and with projections of a population of 8.5 billion in 2025 eventually leveling off at approximately 11.6 billion around 2150, the answer is of significance to people today and of even greater significance for those generations yet to come.[3,4]

Our goal in this chapter is to present an ecological perspective on sustainable development with a focus on terrestrial ecosystems. Our backgrounds are primarily in agroecology and environmental sciences—the more applied aspects of ecology—with experience in managed ecosystems. We recognize the need for a system-wide ecological perspective that focuses on the dynamic nature of complex environmental problems and, in this sense, we see ourselves as the type of ecologists who are having "…a growing involvement in the design and implementation of development projects, as governments move to protect air and water quality, conserve natural resources, and support economic development with sound environmental management."[5] As we consider sustainable development from an ecological perspective, we should have some sense of attributes associated with ecosystems and an appreciation for ecosystem integrity, which is a basic concept in ecological sustainability.

ECOSYSTEM ATTRIBUTES

An ecosystem is more of a concept than a real, physical entity. Tansley first formally proposed the term ecosystem with regard to a single location as "not only the organism complex, but the whole complex of physical factors forming what we call the environment."[6] Another view of an ecosystem is as a system, that is, a collection of interacting components and their interactions that includes ecological or biological components. A key idea is that an ecosystem includes the physical or abiotic environment in addition to the biological components.[7]

Ecosystems have six major, defining attributes:[8]

(1) Structure, which is the composition and arrangement or the distribution of matter and energy among the biotic and abiotic subcomponents

(2) Function, which is the integrated holistic dynamics which result from a constant exchange of matter and energy between the physical environment and the living community

(3) Complexity, which results from a high level of biological integration and may occur at several hierarchical levels

(4) Interaction and interdependency among the various living and nonliving components, which ensure that a change in one component will result in a change in other components

(5) Spatial boundaries and scales that are diffuse and multitiered

(6) Temporal change, which is inherent in ecological systems and can result in changes in an ecosystem's entire structure and function given enough time

ECOSYSTEM INTEGRITY

Ecosystem integrity refers to the soundness or completeness of the system and its existence in a state of being whole and unimpaired. The integrity of both the system structure and function, a maintenance of system components, interactions among them and the resultant dynamic of the ecosystem are implied. However, as King states: "...to a considerable degree, the intuitive concept of ecosystem integrity is biased towards function integrity, the state of being unimpaired."[7]

Additionally, the actual assessment of ecosystem integrity is dependent upon the perspective of the observer. The perspectives which people have in relation to indicators like economics and aesthetics and the perception of intended function of the ecosystem will bias judgments about ecosystem integrity. Human value judgments related to economics or aesthetics should not be excluded in assessments of ecosystem integrity for natural, ecological or scientific perspectives. There must, however, be a balanced approach in judging what constitutes sustainable development and the ecological perspective must play a prominent role in any such assessment.

Temporal and spatial scale are key elements in assessing ecosystem integrity. The temporal scale must be sufficiently long to identify what the "normal" state of the ecosystem should be and to determine whether departures from "normal" are indeed trends or merely random variations. Long-term observations are needed to reveal slow changes in a system component that would be viewed as constant with short-term observations. Similarly, observations made on a large spatial scale may reveal heterogeneity not apparent from limited local observations.

PERSPECTIVES ON SUSTAINABLE DEVELOPMENT

Why Sustain Ecological Resources?

Before examining any perspective on sustainable development, it is important to know what we want to sustain and why we want to sustain it. The answers to these questions are essentially issues of setting goals that we can live with given our personal world view. Also, it is important to recognize the costs of environmental and ecological damages and to inject them into the decision-making process as early as possible.[9] As Noss points out with regard to sustainable forestry, we would have an easier task if our goal was to maintain an approximately even flow of wood products than if we are actually concerned about sustaining food webs and nutrient cycles that maintain soil productivity in forest ecosystems.[2]

Also, from an agroecological perspective, some see sustainability as an extension of current concerns about food security into the future; that is, if food production is maintained or increased, then we have sustainability—this might be considered sustainable development.[10] Whereas it is true that maintenance of production of food is a necessary index of agricultural sustainability, it is also clear that production of commercial food crops alone is not a sufficient index of sustainability for agroecosystems.

Given that people are increasingly the dominant force in ecosystem change, particularly in terrestrial ecosystems, our view is that we should preserve as many intact and naturally functioning ecosystems as possible along with those managed ecosystems essential to the survival and well-being of people. The dilemma, of course, is that there are few, if any, pristine or even only marginally altered ecosystems and that in many areas of the world terrestrial ecosystems have already been transformed largely into managed agricultural, forest and rangeland ecosystems with a primary purpose of providing food, fiber and shelter for people. Survival of native plant, animal and microbe species is of concern in many undeveloped and underdeveloped areas and within nature reserves; however, the reality is that, through what Crosby has called ecological imperialism, many of the temperate lands have already been transformed into "new Europes" and development in environmentally sensitive, tropical areas of the world is continuing at an accelerating rate.[11]

Toman presents two possible views in defining sustainable forestry that can be generalized to help answer these questions of why we wish to sustain ecological resources.[12] One view of sustainability is that all resources—our natural endowment, physical capital, human knowledge and abilities—are relatively fungible sources of well-being. With this view, damage to ecosystems due to factors such as degradation of environmental quality, loss of species diversity or global warming are intrinsically acceptable. Rather, the question becomes whether compensatory investments for future generations in other forms of capital such as human knowledge, technique and social organization are possible and are undertaken.[13]

Another more ecologically based view is that such compensatory investments are often infeasible and are often ethically indefensible when they include ecosystem degradation. Physical laws limit the extent to which resources can substitute for ecological degradation and no practical substitute may be possible for natural life-support systems.[14] Compensatory investments may not be meaningful if ecosystem degradation has progressed to irreversible levels.

Two alternative views of why we may want to preserve biodiversity, which is a component of ecological sustainability, also illustrate diversity of perspectives.[2] One view is that biodiversity provides a genetic warehouse from which we can draw all sorts of useful products. An alternative view is that biodiversity should be preserved, because it is an end in itself. In essence, we are interested in preserving the full spectrum of species, genetic material and ecosystems on earth, because they have an inherent worth that overshadows any use we might conceive for them. The first view supports the concept of sustainable development, whereas the second is directed more at the sustainability of ecosystems.

People and Ecological Sustainability

Two somewhat divergent views can be taken as to the role of people in ecological sustainability. The prevailing view is that sustainability is fundamentally a human construct. Therefore, any discussion of sustainability must be strongly conditioned by human values, whether the discussion emphasizes the short-term utilization of ecological resources, provision for future uses and intergenerational equity or, in the case of species, their inherent right to exist.[9,15] Pfister echoes this sentiment and suggests that the best of ecological approaches can only sustain ecosystems when such approaches are integrated into the human context.[16] Noss agrees that the concept of sustainability has been applied previously in a nearly anthropocentric manner and that it is in that very area the concept is in need of "significant revision."[2]

The second view provides a step toward such a revision in that human society is seen as an integral part of natural ecosystems. This latter view, while laudable and certainly a worthy goal, may be unrealistic given the extent to which people have already changed global ecosystems and the anticipated increase in number of global inhabitants.[4,11] However, the second view may still be a reality, if we can accept the changes already made and move toward human society becoming more of an integral part of natural systems.

Sustainable Development Balances Ecological, Economic and Societal Values

Many viewpoints exist with regard to sustainable development. Such viewpoints or perspectives are not independent but are rather intertwined to provide a particular person's view of sustainable development. The ecological perspective is an essential viewpoint; however, it is only one of several important viewpoints and must be placed in context to be fully understood.

In a survey of definitions for the term "sustainable agriculture," Neher concluded that there were three common themes in such definitions: plant and animal productivity, environmental quality and ecological soundness, and socioeconomic viability.[17] Three definitions of sustainable agriculture illustrate these themes:

...Sustainability refers to the ability of an agro-ecosystem to maintain production through time, in the face of long-term ecological constraints and socioeconomic pressures.[18]

Sustainable agriculture is a complex concept incorporating ecological stability and reliance (e.g., conservation of resources and reduction of impacts on the environment), economic viability, the quality of life, and human welfare.[10]

Sustainable agriculture over the long-term enhances environmental quality and the resource base upon which agriculture depends, provides for basic human food and fiber needs, is economically viable, and enhances the quality of life for farmers and society as a whole.[19]

We believe the third definition best describes our concept of sustainable agroecosystems, but we would add a wildlife component to the definition. We should also acknowledge that agriculture was developed as an enterprise of human activity to even out environmental and economic risk while maintaining a productive base over time.

The ecological perspective is one of three key conceptual issues or sets of objectives with potentially important operational implications for sustainable development (Figure 2.1). Substantial effort has been invested to ascertain the implications inherent in the three components of sustainable development.[20] Ecologists stress preserving the integrity of ecological systems that are critical for the overall stability of our global ecosystem and deal in measurement units of physical, chemical and biological entities.[9] Economists seek to maximize human welfare within the existing capital stock and technologies and use economic units (i.e. money, or perceived value) as a measurement standard. Sociologists emphasize that the key actors in sustainable development are people with a range of needs and desires and use units which are often intangible, such as well-being and social empowerment.

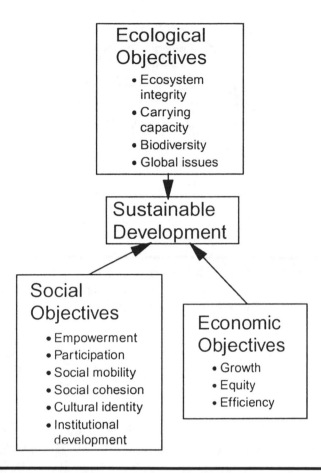

FIGURE 2.1 Three sets of objectives to be integrated for sustainable development. Although the emphasis given to a set of objectives depends on one's viewpoint, all of the objectives must be brought into concordance before sustainable development can be achieved. (After Serageldin.[9])

Sustainable solutions for the development of terrestrial systems fall at the intersection of the spheres that represent the three key ingredients for sustainable development (Figure 2.2). Sustainable development occurs only when management goals and actions are simultaneously ecologically viable, economically feasible, and socially desirable; these imply environmental sound-

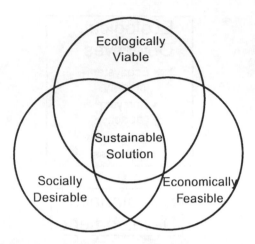

FIGURE 2.2 Sustainable development is achieved though the intersection of the three key elements that encompass the viewpoints of ecologists, economists and sociologists.

ness and political acceptability. The science of ecology itself provides no basis for value judgments; rather it provides a basic source of information upon which socioeconomic decisions can be made, as conditioned by people's values.

However, imbalance among the three components or recasting of the balance to reflect only the primacy of one viewpoint will likely result in failure to achieve sustainability because of failures in one or more of the spheres.[21] For example, if a primarily economic viewpoint is taken which casts economic objectives only as growth and efficiency but reduces ecological objectives solely to natural resource management, a sustainable balance cannot be achieved over the long term.[9] Similarly, sustainability of terrestrial systems (e.g., forests) is intimately wrapped up with human values and institutions, not just ecological functions.[12] Further, what may appear superficially as biophysical and technical problems with resource degradation and loss of potential in systems such as agroecosystems is often in reality rooted in economic, social and political issues.[10]

Environmental or Ecological Sustainability?

Two similar but not wholly equivalent terms are used to refer to the component of sustainability that encompasses ecosystem integrity, biodiversity, and similar ideas. *Environmental sustainability* is used more often in discussions that relate to resource economics and accounting (i.e., sustainable development), and *ecological sustainability* is used more often by biologists in discussions of sustainable forestry or sustainable agriculture. We prefer ecological sustainability, because it retains more of the concepts of connectedness, complexity and diversity of the biological and environmental areas and maintains a point of view that, while somewhat anthropocentric, is not exclusively centered on ecosystems as sources and sinks for human activities.

Environmental sustainability is often used in a narrower context that refers primarily (although not exclusively) to the physical environment and to flows of goods and services from natural capital, where natural capital, according to Goodland and Daly, is "basically our natural environment and is defined as the stock of environmentally provided assets (such as soil, atmosphere, forests, water, wetlands), which provide a flow of useful goods or services." For example, Goodland and Daly state that environmental sustainability "...seeks to improve human welfare by protecting the sources of raw materials used for human needs and ensuring that the sinks of human waste are not exceeded, in order to prevent harm to humans."[22]

SCALES OF ECOLOGICAL SUSTAINABILITY

Temporal and Spatial Scales

Scales provide an essential frame of reference against which to measure ecological sustainability. In part, we ask again, "What is it we wish to sustain" but we add important modifiers such as over what time and what area and in what location in the world.

The temporal scale of interest is probably relatively short when viewed in relation to geological time but quite long when viewed in relation to the lifetimes of people. The time scale of proximate interest is thus probably hundreds of years to a millennium, as adopted by Daily and Ehrlich in considering population, sustainability and earth's carrying capacity.[3]

The upper limits of spatial scale are constrained, with current technology, to the size of our planet. For most terrestrial systems, spatial scales of interest are at a continental scale or often at smaller areas of geopolitical units or landscape types. Other spatial units of interest may be watersheds, ecoregions or major land resource regions.[23,24] Spatial scales are extremely important in the development of systems for sustainable management of ecological resources.[25,26] The spatial scale of predominant interest will be dependent on the environmental and biological processes of interest. For example, an examination of the effects of ambient concentrations of ozone or other pollutants on forest or agricultural ecosystems will understandably have a larger spatial scale of interest than will an evaluation of how tillage practices affect soil communities and soil structure.

Locational perspectives are also of interest in examining issues of ecological sustainability. For agricultural systems in the developed world, for example, where vast alterations of the landscape and environment have already occurred and many temperate ecosystems have a high degree of resilience, the main issues of sustainability include diversification from a relatively limited range of commodities and reduction of flows of nutrients and pesticides from agricultural systems into adjacent systems.[10] For agricultural systems in developing countries, the issues of diversification are not as important as are fundamental issues of food production. Also, because of the fragility of many tropical ecosystems, the impetus must be to produce food in increasing amount without destroying the ecological base upon which plant growth is dependent.

Role of Landscape Ecology

Land use is a vital factor in the ecological sustainability of resource development, and important aspects of land use, from an ecological perspective, are the arrangement, size and connectivity of landscape elements. If we consider habitat for larger animals, then habitat patch size, edge-to-area ratios and the degree of fragmentation or connectivity of suitable habitat units are of primary importance in determining overall habitat suitability. Also, the relative partitioning of remnant forest patches, buffer zones and windbreaks in relation to streams and riparian areas can be quite important for sustaining biodiversity as well as for resource preservation. Landscape fragmentation, for example, is currently the major problem affecting the long-term sustainability of the central hardwood forests in the United States.[27]

The rapidly developing discipline of landscape ecology offers considerable promise for documenting and interpreting the dynamics of landscapes, evaluating land suitability and integrating ecological, social and economic precepts for ecosystem development.[16] Currently, our scientific understanding of how landscape design can be used to maintain long-term biodiversity in the range of terrestrial ecosystems is incomplete;[28] however, scientific inroads in specific areas have been encouraging.[29,30]

Studies in landscape ecology provide an opportunity to view the possibilities of ecologically sustainable development at larger and more meaningful spatial scales—landscapes and regions. It is essential that we move beyond a concentration on individual landscape patches to examining the interaction among various types of landscape patches with other landscape elements (e.g., lakes, streams). Landscapes are perhaps the optimal scale for planning and management associated with sustainable development or a sustainable environment.[31] Landscape ecology, as it develops further as a science, will allow us to assess not only the ecological sustainability of ecosystems but landscape as well, which will then provide guidance for landscape-level management for sustainability.

MONITORING ECOLOGICALLY SUSTAINABLE DEVELOPMENT AND ECOSYSTEM HEALTH

In order to ensure that ecological sustainability remains a reasonably balanced component of the three spheres of interest (Figure 2.2) so that solutions for sustainable development can be derived, it is essential that quantitative, scientifically sound measures of ecological sustainability or ecosystem health be available. Currently, the quality of information and comparability of information across even fairly small spatial scales (i.e., watershed), with regard to ecological or environmentally sustainable development, remains poor.[32] The lack of indicators of ecological condition relative to economic condition, for example, is primarily because it is easier to measure transactions in the marketplace than to measure biotic and abiotic interactions among ecosystem components. Also, obtaining an inventory of economic capital is relatively easier than obtaining such an inventory for biological or ecological capital.

One attempt to quantify and monitor the relative sustainability of development and health of U.S. ecological resources is the Environmental Monitoring and Assessment Program, an interagency and interdisciplinary program originated by the U.S. Environmental Protection Agency.

The Environmental Monitoring and Assessment Program

The primary goal of the Environmental Monitoring and Assessment Program (EMAP) is to track the condition of the ecological resources in the United States. EMAP is concentrating on ecological sustainability.

EMAP was started in 1989 to address the following questions: "Do environmental policies work?" and "What new ones do we need?" To answer these questions, it is necessary to understand the extent and severity of environmental problems. Other monitoring efforts are/were site- or problem-specific and did not allow assessment of condition over any spatial scale with statistical confidence. EMAP is designed to collect relevant information to help policymakers decide how to allocate limited resources among different environmental problems. The effort has expanded beyond the EPA to include other federal agencies, universities and scientists from a wide variety of disciplines.

EMAP is developing a set of indicators that will describe the condition of ecological resources. It is not designed to address all ecological issues, but to provide statistically representative information about the condition of a region. Monitoring will take place at multiple scales; information collected at smaller scales is primarily for integration into regional- and national-level assessments (Figure 2.3). Site-specific problems will generally be overlooked and may need to be addressed by other site-specific efforts and programs. However, the indicators used by EMAP will be valuable in placing local conditions in a larger geographic context. EMAP information will be provided to resource managers and decision-makers.

EMAP has research, monitoring and assessment components that are interrelated. A comprehensive monitoring and assessment program such as EMAP is unprecedented. Research is driven by concepts and ideas from earlier assessment projects, as well as from environmental sciences, and the needs of policymakers and other users. EMAP is not intended to undertake site-specific research. The major environmental issues of the present and the future are issues that cross resources and regions. Existing state and federal monitoring networks, by themselves, typically have not been designed to be anticipatory, to permit multiple resource analyses or to support regional-scale assessments; EMAP was created to fill that gap. The monitoring component is shaped by the methods, strategies and ideas developed in the research component.

Assessment is the key to EMAP. New and innovative approaches are being considered to define current conditions and to assess future change in ecological resources. The integration

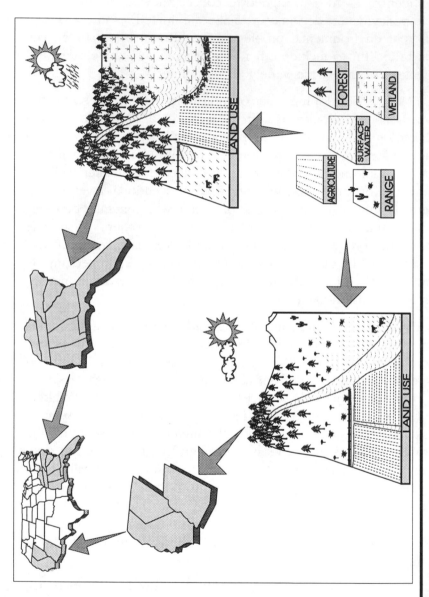

FIGURE 2.3   Monitoring the condition of ecological resources occurs at multiple scales, with information collected at smaller scales being integrated to provide regional- and national-level assessments.

of indicator measurements, conceptual models and supporting data, as developed by EMAP, will provide a report card on the cumulative effectiveness of U.S. environmental policies and will form a foundation for future resource management and policy decisions.

The Terrestrial Component of EMAP

Terrestrial resources are continuous and require an integrated, holistic approach for monitoring and assessment of condition. To date, the principal focus of EMAP's Terrestrial Resource Groups (Agricultural Lands, Rangelands, and Forests) has been to develop indicators for monitoring, assessing and reporting on the ecological condition of terrestrial resources in the United States. Results will provide integrated information and assessments at regional and national scales, over time, in a form that will be useful to public officials responsible for managing terrestrial resources and establishing environmental protection policies. The design and implementation of the Terrestrial Program is guided by a number of principles:

(1) The program emphasizes joint studies among Terrestrial Resource Groups on indicator development, analytical methods, assessment techniques and other technical issues.

(2) Ecological condition estimates are based on probability samples.

(3) Existing information and data bases are networked into the Terrestrial Program, wherever possible.

(4) Landscape features are an integral part of the condition of terrestrial resources.

CONCLUSIONS

Ecological sustainability must not be the driving component but must, at least, be an equal component in continuing discussions on sustainable development. People occupy a central position in any discussion of sustainable development and are the instruments and beneficiaries, as well as the victims, of any development activity.[9] People will continue to be the focus of discussions on sustainable development. However, the rapid depletion of our natural resources, coupled with the degradation of land, air and water quality, indicate that the human global economy, if it continues along its current course, will severely limit the human carrying capacity of our planet.[3] Our global ecosystem, the source of all the physical, chemical and biological resources needed to support our economic and sociological systems, is finite and has now reached a stage where its regenerative and assimilative capacities have become strained.[22]

Additional knowledge is needed on the structure and function of terrestrial ecosystems, their resilience and sensitivities, and their interactions to form landscapes. Acceptance among scientists and policymakers that managed ecosystems are the way of the future is essential, and ways must be found to evaluate and monitor ecosystem health at many spatial and temporal scales. Such information on ecosystem health will not be obtained easily or inexpensively; however, its availability is essential.

Rees maintains that "ecologists will be effective in influencing policy only if they can demonstrate that sound, ecological policies will promote, not hinder, sustainable development."[5] We find this view to be acceptable in the short term, but totally unacceptable in the long term. Ecological sustainability must receive consideration at least equal to that of economic sustainability. Ecologists also must assume a larger role in policymaking itself, and in this we agree with Noss that "environmental policy is too important to be left to the policymakers, most of whom know little and care little about forests or sustainability in the broadest ecological sense."[2]

REFERENCES

1. G.H. Brundtland, Chair, *Our Common Future,* Oxford University Press, New York, 1987.

2. R.F. Noss, in *Defining Sustainable Forestry,* G.H. Aplet, N. Johnson, J.T. Olson and V.A. Sample, Eds., Island Press, Washington, D.C., 1993, pp. 17–43.

3. G.C. Daily and P.R. Ehrlich, "Population, sustainability, and Earth's carrying capacity," *BioScience,* 42:761–771, 1992.

4. United Nations Population Fund (UNFPA), *The State of World Population 1991,* UNFPA, New York, 1991.

5. C. Rees, "The ecologist's approach to sustainable development," *Finance & Dev.,* 30:14–15, 1993.

6. A.G. Tansley, "The use and abuse of vegetational concepts and terms," *Ecology,* 16:284–307, 1935.

7. A.W. King, in *Ecological Integrity and the Management of Ecosystems,* S. Woodley, J. Kay and G. Francis, Eds., St. Lucie Press, Delray Beach, FL, 1993, pp. 19–45.

8. J.P. Kimmins, *Forest Ecology,* Macmillan, New York, 1987.

9. I. Serageldin, "Making development sustainable," *Finance & Dev.,* 30:6–10, 1993.

10. M.J. Jones, in *Crop Protection and Sustainable Agriculture,* D.J. Chadwick and J. March, Eds., Wiley, Chichester, 1993, pp. 30–47.

11. A.W. Crosby, *Ecological Imperialism: The Biological Expansion of Europe, 900–1900,* Cambridge University Press, Cambridge, UK, 1986, pp. 1–7, 145–170, 171–194, 295–308.

12. M.A. Toman, in *Defining Sustainable Forestry,* G.H. Aplet, N. Johnson, J.T. Olson and V.A. Sample, Eds., Island Press, Washington, D.C., 1993, pp. 270–279.

13. J.M. Hartwick, "Intergenerational equity and the investing of rents from exhaustible resources," *Am. Econ. Rev.,* 67:972–974, 1977.

14. C. Perrings, "Conservation of mass and instability in a dynamic economy–environment system," *J. Environ. Econ. Manage.* 13:199–211, 1986.

15. J.T. Franklin, in *Defining Sustainable Forestry,* G.H. Aplet, N. Johnson, J.T. Olson and V.A. Sample, Eds., Island Press, Washington, D.C., 1993, pp. 127–144.

16. R.D. Pfister, in *Defining Sustainable Forestry,* G.H. Aplet, N. Johnson, J.T. Olson and V.A. Sample, Eds., Island Press, Washington, D.C., 1993, pp. 217–239.

17. D.A. Neher, in *Integrating Sustainable Agriculture, Ecology, and Environmental Policy,* R.K. Olson, Ed., Haworth Press, New York, 1993, pp. 51–61.

18. M.A. Altieri, *Agroecology: The Scientific Basis of Alternative Agriculture,* Westview Press, Boulder, CO, 1987.

19. N. Shaller, "Mainstreaming low-input agriculture," *J. Soil Water Conserv.,* 45:9–12, 1990.

20. World Bank, *World Development Report 1992: Development and Environment,* Oxford University Press, New York, 1992.

21. I.S. Zonneveld, in *Changing Landscapes: An Ecological Perspective,* I.S. Zonneveld and R.T.T. Forman, Eds., Springer-Verlag, New York, 1990, pp. 3–20.

22. R. Goodland and H. Daly, "Environmental sustainability: universal and non-negotiable," in *Ecological Economics: Building a New Paradigm for Sustainability,* Ecological Society of America, Knoxville, TN, 1994.

23. J.M. Omernik, "Ecoregions of the conterminous United States," *Ann. Assoc. Am. Geographers,* 77:118–125, 1992.

24. U.S. Department of Agriculture–Soil Conservation Service, Land Resource Regions and Major Land Resource Areas of the United States, Agric. Handbook 196, USDA-SCS, Washington, D.C., 1981.

25. J.A. Wiens, "Spatial scaling in ecology," *Funct. Ecol.,* 3:385–386, 394, 1989.

26. D.L. Urban, R.V. O'Neill and H.H. Shugart, Jr., "Landscape ecology," *Bioscience,* 37:119–127, 1987.

27. G.R. Parker, in *Defining Sustainable Forestry,* G.H. Aplet, N. Johnson, J.T. Olson and V.A. Sample, Eds., Island Press, Washington, D.C., 1993, pp. 202–216.

28. M.G. Turner, "Landscape ecology: the effect of pattern on process," *Annu. Rev. Ecol. & Syst.,* 20:171–197, 1989.

29. E. Grumbine, "Protecting biodiversity through the greater ecosystem concept," *Natural Areas J.,* 10:114–120, 1990.

30. A.J. Hansen, T.A. Spies, F.J. Swanson and J.L. Ohmann, "Conserving biodiversity in managed forests," *Bioscience,* 41:382–392, 1991.

31. R.T. Forman, in *Changing Landscapes: An Ecological Perspective,* I.S. Zonneveld and R.T.T. Forman, Eds., Springer-Verlag, New York, 1990, pp. 261–280.

32. A. Steer and E. Lutz, "Measuring environmentally sustainable development," *Finance & Dev.,* 30:20–23, 1993.

3

AFTER RIO: THE NEW ENVIRONMENTAL CHALLENGE

Si Duk Lee and Victor S. Lee

Simply based on the number of world leaders who attended, the United Nations Conference on Environment and Development (UNCED), commonly known as the Rio Conference, was heralded as a great achievement even as it occurred. Afterwards, commentators again trumpeted the great success of the conference because of the so-called historic documents that were produced by the attendees. In addition, everyone agreed upon the necessity for sustainable development.

"Sustainable development" has had widespread political usage because of its broad application and vague definition. In many circles, the term stands for "environmentally compatible development." UNCED, held in June 1992 in Rio de Janeiro, Brazil, was the first widely publicized recognition that environmental quality and economic health are inextricably linked. The politicians and

Dr. Si Duk Lee is a researcher in environmental risk assessment and health effects. He has also served as the Chairperson for the International Affairs Committee of the Air and Waste Management Association. Victor Lee is the Director of the Center for Environment and Development, P.O. Box 13661, Research Triangle Park, NC 27502.

governmental officials who attended emerged with a broad consensus for the necessity of sustainable development; however, they did so without agreeing on its meaning.

Sustainable development is a term that has been ambiguous and confusing in its operational definition. It has been well observed that the term has taken on whatever meaning speakers have needed to promote their cause. Some have cynically suggested that the ambiguity of the term may be what has made it so widely acceptable and that agreement on an undefined goal can obscure underlying differences.

The first use of the term was purely economic—sustainable development strictly meant keeping business growing and moving forward. The term has slowly evolved into a combination of environmental conservation and economic growth, but everyone has retained his own definition. A significant step for the environmental movement took place when the World Commission on Environment and Development submitted its final report, titled *Our Common Future,* in 1987 to the United Nations General Assembly and proposed a definition of sustainable development that became a general starting point for worldwide agreement.[1]

Our Common Future, considered to be the bible for sustainable development, states the principle as the merger of environmental and economic concerns: the environment cannot be effectively protected without economic development, and economic development cannot be sustained without environmental protection. This definition was part of the result of a long and arduous journey toward an educated perspective on environmental conservation.

In fact, there has been tremendous effort expended to simply find an agreeable definition for sustainable development. Part of the problem for too long has been that the goals of business leaders and environmentalists were polar opposites. In addition, a continuing problem has been that those within these polarized groups have not themselves agreed on a common goal.

It has been, and continues to be, a sad chapter in world history that so-called leaders are bickering over definitions instead of actually doing something about cleaning up the environment. A common understanding of sustainable development will not eliminate the need to try to resolve competing values, but it may produce consensus on the need to try for balance. However, it is certain that finding an agreeable definition will not protect and improve the environment one bit.

The fact is that there is too much posturing and not enough action. Many leaders of different countries have shown no serious interest in improving the environment; rather, they are using it as an excuse to obtain an economic advantage. From the proceedings of recent world gatherings, it is quite evident that we have a long way to go toward the full understanding of effective environmental conservation.

REFLECTING ON RIO

UNCED was a prime example of effective posturing and ineffective action. UNCED produced three documents which were seen at the time as major advances towards achieving international cooperation for environmentally sustainable development. These documents were the Rio Declaration on Environment and Development, an "action plan" called Agenda 21 and a statement of sustainable management principles for the world's forests.

Throughout the three documents, sustainable development was the underlying theme. Twelve of the first twenty-seven principles of the Rio Declaration have sustainable development as a primary focus. But the documents were all nonbinding. This gave many countries the opportunity to show the world that they were taking a stand for environmental concerns, but they committed themselves to no particular action. However, since no country would have to follow through on its commitment, the document signings were, in many ways, a sham. This high-profile circus received tremendous media attention, whereas little publicity was given to the tons of garbage that descended upon a country, Brazil, which was inadequately prepared for its disposal. It was a sad sign of ineffectiveness that a world bureaucracy brought together to clean up the environment could so effectively pollute it.

This disheartening depiction of UNCED was well detailed in a recent news article:

> Words, words, words. Shakespeare's description seemed all too applicable to the second meeting of CSD [Commission on Sustainable Development] in New York. Disappointingly few decisions were made. Preparations for the next meeting, in 1995, will have to be conducted more thoroughly and at a high level. [Nitin] Desai explained at the press conference after the New York session: "Most national governments were represented here by their environment ministers. They are important, but do not usually have enough political weight to ensure that all the problems are ironed out. We will have to involve ministers responsible for planning, finance and economic affairs at an earlier stage next time. This new UN agency will only be able to take decisions if there are concrete plans on the table to which ministers can say yes or no." The length of the document signed at Rio shows that the implementation of Agenda 21 will require an enormous number of plans. The CSD is responsible for coordinating and streamlining, and that will be a hell of a job. This is regarded as one of main reasons why the Commission is making so little headway.[2]

A highly disturbing result of the Rio Conference was a rise in the "North–South" tensions—the competing interests between developed and developing countries. These conflicts were extremely heated and focused heavily on two issues: financial resources and technology transfer. These disagreements seemed headed for an impasse, but workable resolutions were reached by the end of the conference.[3]

The developing country bloc, calling itself the G-77, insisted on commitments from the North for additional financial resources to implement Agenda 21 and commitments to assist the South to obtain the necessary technologies. At the same time that this demand was being made, the G-77 resisted committing to a program that specified how the technologies would be used or deployed and also resisted committing to actually spending funds received for environmental improvements.

Obviously such a situation was ludicrous for countries of the North to accept, especially when many of the G-77 had a history of abusing financial aid in the past or had no structure presently to use the financial resources given to them. The G-77 resisted the North's insistence on such commitments on the basis that additional conditions applied to financial assistance would be inappropriate and would be an infringement on their sovereignty.

This general tension spilled over into other issues and increased the tension as negotiations proceeded. For example, population growth concerned many countries. However, most countries resisted a high priority for this issue due to the resistance of the countries of the North, the United States in particular, to the inclusion of a chapter in Agenda 21 dealing with "consumption patterns." This term refers to the pattern of production and consumption of goods and services found mainly in the countries of the North. Eventually Agenda 21 included chapters on both population and consumption patterns, but the United States strongly resisted including a chapter on consumption patterns and stirred up extensive resentment among developing countries.

From the start, developing countries expressed unhappiness with the "Northern-driven" agenda of UNCED and the preoccupation of developed countries with global issues. The G-77 frequently interjected the proceedings with their own concerns with poverty and the need for economic growth and development to counter the agenda of the developed countries. The G-77 also complained that the developed countries interjected the "global concerns mentality" into debates only where it was to their own benefit and where those concerns more deeply affected the interests of the underdeveloped countries.

UNCED was the first step on the way to environmental cooperation, but there is certainly a long way to go. The UNCED documents contained only nonbinding recommendations or principles which facilitated agreement and cooperation among the participants. It should be noted, however, that the wording was carefully negotiated in the three major documents since it is expected that the agreed-upon language in the documents will affect how these issues are framed in future forums.

At this time, it appears that the actions of the attendees were only intended to establish commissions and official positions regarding the environment. The Commission of Sustainable Development (CSD) was established by the U.N. General Assembly in the fall 1992 session as a result of the recommendations of the chapter in Agenda 21 entitled "Institutional Arrangements." Other commissions for sustainable development were formed by the European Community, Japan and the United States. No significant environmental action has been taken by any of these bodies as of this writing.

WHAT NOW?

The Need for Action

What is clear from UNCED and other smaller conferences is that there is a lack of cooperation and a lack of the necessary spirit that will get things done. It is not an original observation that everyone involved is out for his own benefit. In some cases, a clean environment is possible as a byproduct of other objectives, but it is also certain that some countries have objectives that do not take a cleaner environment into account at all.

It is understandable that developing countries want to improve their economic status, but it is also reasonable for developed countries to be hesitant to bankroll another country's growth. These are the simple positions of the North and South countries, yet the opposing sides refuse to recognize the legitimacy of each other's views. As much as idealists would like everyone to cooperate for the good of the earth, it is undeniable that economics and environmental protection are too tightly interwoven to try and ignore one or the other.

What is clear is that nations must be made to *commit* to environmental conservation and stop posturing for pure political and financial gain. The Rio Conference was a good first step, since world leaders gathered in the name of environmental concerns, but it was essentially a tiny first step.

Environmental awareness was raised around the world by the sheer magnitude of the event, but what kind of durability will this awareness have since the spotlight has been turned off? It is now quite evident that words still have not turned into actions. The longer no action is taken, the greater the chance that the small momentum gained from the Rio Conference will continue to ebb away. Time is wasting away as our commissions and conferences argue over the problems and only agree to argue more at the next meeting.

Is there anything that can be done to end the talk and begin action? It seems that any solution requires more bureaucracy and more old ideas that have not worked in the past. Is it possible to even get past the rhetoric before plans are implemented? After all, no one is going to start pouring resources into environmental efforts without having a large say in how these resources will be spent and who will benefit.

There is no single solution. The answer lies in a multifaceted approach that involves the participation of everyone and also does not place the burden on any one group or country. The first step is leadership. So far, it has been clear that governments are exceptional at running conferences, establishing commissions and setting agendas. It has also been clear that these same governments are rather unsuited to carry out the proper course with practical plans and procedures.

A straightforward approach would be to have the United Nations or any other combination of countries arrange meetings to address the pertinent issues. However, the attendees should include more experts and specialists in their respective fields, rather than politicians and government bureaucrats. In this situation, attendees could spend more time addressing the issues and conducting a real exchange of ideas, instead of lecturing officials with little or no background in these areas.[4]

In fact, the involvement of many political figures in the environmental movement has been an insult to the scientific community and the true environmentalists. Since the environmental movement has become so popular, many politicians have made a superficial study of the subject. Afterwards, feeling they are experts, they report their "findings."

Politically motivated acts like this do a disservice to people, too numerous to count, who have devoted not only years but their lives to environmental conservation. It is these scientists, the true environmentalists, who must be brought in and interwoven into the process of putting the world onto the correct path.

But the real question of leadership is still open. At this time, the answer lies in the corporate world. In cases in which governments have enacted strict regulations for polluting, business has responded. But it is generally agreed that this "command and control" approach only goes so far. It has been proven time and time again that regulation and legislation result in reluctant compliance rather than an orientation of innovation and continuous improvement.

Private Sector Initiatives

Many feel that the "reward" method—giving incentives to business to comply instead of merely punishing it when it does not comply—and market-based incentives achieve better results and more rapid responses from the corporate world. Effective sustainability requires building an economic advantage after finding the costs and benefits of environmental improvements.[5]

However, some of the most interesting and innovative environmental programs have come when a business has taken it upon itself to conserve and protect the environment. There are many examples in which companies have made environmental conservation changes for substantial economic savings.[6]

An example of the benefits of general conservation came during the 1970s. While it had been a widely held assumption that energy consumption must grow proportionately with economic growth, U.S. energy consumption remained constant after the "energy crisis" ended as U.S. economic output increased by more than one-third.

Since 1986, energy growth has again mirrored economic output, but it was proven that the two are not inevitably intertwined. Now in the 1990s, it is even clearer that there are substantial opportunities for savings from separating energy use and economic growth through investments in energy-efficient sources and methods. It is up to the corporate world to take it upon itself to search for these innovative solutions again.

An example of a corporation becoming serious about environmental conservation is the 3P Program of the Minnesota Mining and Manufacturing (3M) Corporation. 3P stands for Pollution Prevention Pays. The idea for the program was inspired by the environmental awareness of the 1970s. The corporate leadership saw the effects of new regulations and laws on bottom-line profits and renewed efforts to reduce waste and pollutant releases. The goal of their program, which is now the much-imitated model, was to both comply with regulations and reduce the costs of pollution control.

The idea of pollution avoidance was a change from the standard practice of installing "end-of-pipe" pollution controls. The cost for such "add-on" controls was very high and, in reality, provided only a temporary solution. A more proactive philosophy was to search for more efficient and durable solutions.

This philosophy of environmental conservation evolved into an environmental policy and took 3M out of a reaction mentality and into an action mentality. This philosophy includes the following guiding principles:

- Solve our own environmental pollution and conservation problems;

- Prevent pollution at the source wherever and whenever possible;

- Develop products that will have a minimum effect upon the environment;

- Conserve natural resources through the use of reclamation and other appropriate methods;

- Assure that our facilities and products meet and sustain the regulations of all federal, state and local government agencies; and

- When possible, assist governmental agencies and other official organizations engaged in environmental activities

Within the environmental policy and the 3P Program, practical developments and new goals continue to evolve. For example, 3M has companion programs in recycling, van pooling and energy conservation. 3M has demonstrated that once a policy is put in place, and innovation in environmental conservation is encouraged among all employees, then the very way that business is done changes and becomes more energy and cost efficient.

Since 1975, when the program was started, the worldwide results of 3P have been estimated to be:

- 1.3 billion pounds of releases prevented to air, water and land
- $710 million in savings
- 4,200 projects in 22 countries

This type of organization-wide commitment to environmental conservation shows the depth of dedication that is essential to any program. In the past, the catchphrase "going green" usually meant cleaning up after the fact to avoid fines. But it has become very evident that regardless of restrictions and regulations, raised environmental consciousness, and the heightened level of innovation that follows, can make businesses more efficient and reduce costs. Waste minimization is just one example of the many ways a company can make money by conserving.

Waste Minimization

Waste minimization has been a successful procedure for many organizations. The goals and benefits of waste minimization are to:

- Save money by reducing waste treatment and disposal costs, raw material purchases and other related operating costs
- Reduce potential environmental liability problems
- Protect public health and worker safety
- Conserve the environment

Effective waste minimization programs have several attributes in common, including:

- Top management support
- Explicit program scope and objectives

- Accurate waste accounting

- Accurate cost accounting

- Organization-wide waste minimization philosophy

The two broad categories of waste minimization techniques are source reduction and recycling, both onsite and offsite.

Source reduction encompasses both product changes and source control. Product changes can include product substitution, product conservation, change in product composition and change in product manufacture. Source control offers a diversity of ways to minimize waste:

- Input material changes, such as material purification and material substitution

- Technology changes, such as process changes, equipment or layout changes, added automation and changes in operational settings

- Good operating practices, such as procedural measures, loss prevention, management practices, waste stream segregation, materials handling improvements and improved production scheduling

Recycling can be done onsite and offsite. A combination benefits the environment and company even more. Although definitions vary, we use the following definitions. A product is considered to be "recycled" if it is used, reused or reclaimed. A material is "used" or "reused" if it is either (1) employed as an ingredient to make a product (including use as an intermediate) or (2) employed in a particular function as an effective substitute for a commercial product. A material is "reclaimed" if it is processed to recover a useful product or if it is regenerated. Examples of the latter are the recovery of lead from spent batteries and the regeneration of spent solvents.

Practical onsite use and reuse require materials that are more efficiently used and that minimize waste. Offsite recycling can be aided when a company assists in processing its waste into a byproduct.

General education of this nature teaches the benefits of environmental conservation. It is this type of education and awareness that is strongly needed, not the rhetoric of politicians. Real progress can be made and words can be put into action if we can end the rhetoric and search for real solutions.

CONCLUSION: EFFECTIVE ACTION

Environmental action means real cost savings and actual money-making propositions for business. Estimates for the worldwide environmental market by the year 2000 conservatively range from U.S. $300 billion to as high as $500 billion. Clearly, these are numbers that companies and governments cannot ignore.

As can be seen from the example of 3M, the realm of specialists and experts is not limited to science. There are practical models, such as this, that have been in practice for decades now, and there is much to be learned from their examples. The time for words is long past. It is time to become active and to urge corporations to stop trying to fix problems after the fact and to come up with effective solutions to reduce pollution and eliminate waste.

The single most important factor for more effectively achieving our goal of sustainable development is quality information exchange. This includes a careful study of environmental risk management. In the absence of historical data and information on emerging technologies, it would not be possible to make correct decisions to move toward our goals in a most cost-effective manner.

There is an old saying that those who do not study history are doomed to repeat its mistakes. Although this saying was most often applied to warfare, it is just as true for environmental history. In fact, the environmental movement is a fight of a different kind. Tremendous changes, benefits and savings can be achieved collectively. It is the push–pull of politics in the environmental arena and the "me first" attitude of many countries that will delay those achievements.

NOTES AND REFERENCES

1. World Commission on Environment and Development, *Our Common Future (The Bruntland Report)*, Oxford University Press, Oxford, 1987.

2. *Environmental News from the Netherlands*, 3:11, 1994.

3. The issue of competing North–South concerns is complicated and endless. For a source of discussion on this issue, see Patti L. Petesch, *North–South Environmental Strategies, Costs and Bargains*, Overseas Development Council, Washington, D.C., 1992.

4. There is an interesting discussion of the effective use of independent panels and commissions to investigate environmental policy issues in *World Development Report 1992, Development and the Environment*, Oxford University Press, New York, 1992, p. 88.

5. See Paul Portney, Ed., *Public Policies for Environmental Protection*, Resources for the Future, Washington, D.C., 1990, Chapter 1.

6. For a lengthy discussion of various aspects of corporate environmental issues and considerations, see Bruce Smart, Ed., *Beyond Compliance: A New Industry View of the Environment*, World Resources Institute, Washington, D.C., April 1992.

4

TECHNOLOGIES FOR SUSTAINABLE DEVELOPMENT

Thomas T. Shen

Few would disagree that economic development and environmental protection will top the national and international agendas in the twenty-first century. The United Nations Conference on Environment and Development, which was held in Rio de Janeiro in June 1992, was an important beginning for the process of providing a new environmental agenda based on the concept of sustainable development. The need now is to translate that broad sustainable strategy into specific national action plans as described in Agenda 21 and also to redefine policy and programs for economic development and environmental protection. For technologies, this means profound change— change in thinking and action.

Nations measure the monetary value of goods and services from economic activity as an indicator of national well-being.

Thomas T. Shen presently consults and lectures and is also the author of *Industrial Pollution Prevention.* He held a career position as Senior Scientist with the New York State Department of Environmental Conservation, where he specialized in air pollution control, waste management and multimedia pollution prevention. He is the recipient of numerous awards and is a Diplomate of the American Association of Environmental Engineers. Dr. Shen received his doctorate in environmental engineering from Rensselaer Polytechnic Institute.

Current accounting systems used to estimate productivity do not reflect depletion or degradation of natural resources used to produce goods and services. During much of the last century, both producers and consumers depleted natural resources with little thought for the environmental damage they were causing. We continued to overlook environmental damages until polluted land, water and air began to threaten human health and until native species and ecosystems began to disappear. Environmental–economic interactions in environmental accounting can be best illustrated as in Figure 4.1.

The problems of ever-increasing waste quantity and toxicity have far exceeded our ability to manage environmental pollution by using waste treatment and disposal technologies. If we con-

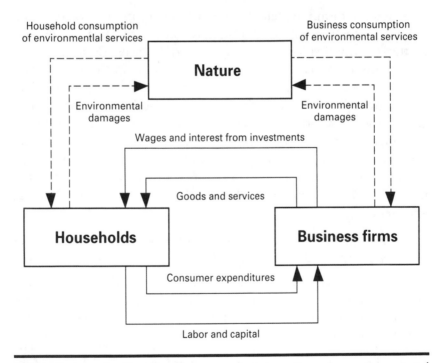

FIGURE 4.1 Environmental–economic interactions in environmental accounting. (Source: Environmental Protection Agency, Washington, D.C., 1992.)

tinue to react to environmental damages and try to repair them, the quality of our environment will continue to deteriorate, and eventually our economy will decline as well. However, if we pioneer new sustainable technologies, shift our policies, make bold economic changes and embrace a new ethic of environmentally responsible behavior, it is far more likely that the coming years will bring a higher quality of life, a healthier environment and a more vibrant economy for all.

This chapter presents some characteristics and the scope of sustainable technologies. It also addresses the need for and the state of the art of new technologies in the realms of energy generation, industrial production, agriculture and transportation because these activities currently account for so much environmental degradation. Topical discussions include design for the environment, the roles of designers and governments in sustainable development and the importance of information and communication. The chapter also provides some reading suggestions.

CHARACTERISTICS OF SUSTAINABLE TECHNOLOGY

The application of technology has been the central means of greater human productivity and, consequently, increases in standards of living. We need new technologies, from materials to processes, that we have not even dreamed of yet. Technology focused on sustainable development is a key to solving problems created in the past and to preventing new ones in the future.

Sustainable technology focuses on pollution prevention and clean technology. Pollution prevention minimizes undesirable effluents, emissions and wastes from products and processes that obviate the need for treatment and control. A preventive approach includes using fewer or nonpolluting materials, designing processes that minimize waste products and pollutants and directing the latter to other useful purposes, and creating recyclable products. Clean technology uses less fuel or alternative fuels to produce energy and generates little or no waste for industry, agriculture and transportation.

Sustainable technologies are those that can reduce environmental pollution through significant technical advances. Because society as a whole benefits from sustainable technologies, they represent appropriate targets for public investment. Technological developments can be considered environmentally and economically sustainable if they:

(1) Reduce environmental pollution cost effectively

(2) Embody a significant technical advance with less waste generation

(3) Are generically applicable at the precompetitive stage

(4) Result in a favorable ratio of social to private returns

The first criterion, environmental pollution, must be interpreted broadly to protect human health, public welfare and ecology. Technological developments must avert serious pollution or markedly reduce the costs of preventing and controlling such pollution.

The second criterion focuses on technologies of the future, which does not imply that today's control technologies cannot yield major environmental improvements. A significant technical advance might well materialize in an entirely new approach to a problem, such as renewable energy sources that obviate the need for fossil fuels.

The third criterion characterizes sustainable technologies in policy terms through public support for private sector technology development. A generic technology is one likely to have wide importance across a class of problems or industrial contexts. Its realization may underlie or make possible the solution to a sequence of technical problems.

The last criterion, which defines sustainable technology by a high ratio of social to private returns, corresponds to a large difference between socioeconomic and financial benefits. Often, research shows, important technological innovations yield greater social and economic benefits than their developers can receive. In addition, market or institutional barriers in some instances make it difficult or impossible for private actors to recoup research and development (R&D) investments in new technology.[1]

In summary, sustainable development technologies are those that depend upon preventive and clean technology. These technologies must be commercially available, economically compatible and environmentally and socially acceptable. Since technology is the key for sustainable development, more resources must be allocated for research and development of new sustainable technologies.

ENERGY TECHNOLOGIES

The energy needs of developing nations will grow at a faster rate than those of industrialized societies. Because 95% of the world's projected population growth will occur in these nations, economic development is essential to their survival. Yet unless more cost-effective energy alternatives become available, wood, coal and oil will be the primary fuels used by developing nations, and traditional energy technologies will be the primary technologies for their economic growth.

Energy extraction, processing and use represent what is arguably the single gravest environmental challenge of the coming decades. Worldwide economic reliance on the conversion of fossil-fuel sources is at the root of the environmental pollution problem. Among various energy technologies, those for producing and using nonfossil-fuel energy sources offer the largest potential to reduce environmental pollution. Many nonfossil fuels and renewable technologies, now in early stages of development, would yield large social returns from technical advances. Technological improvement to yield greater energy efficiency affects the extraction, processing and end use of energy; the potential for further gains is far greater on the demand side than on the supply side. To improve energy efficiency means to reduce the wasteful use of energy, without necessarily entailing any cut in comfort or production. Much of the improvement has been realized by quite simple measures.

One of the greatest technological challenges for the future will be to develop less environmentally damaging sources of energy while simultaneously reducing total energy consumption through better energy efficiency in industry, agriculture, consumer products, the home and in almost every other sector of the economy. New energy technologies will give us a competitive edge as well as a healthier environment. The worldwide market for such technologies will continue to grow as the connections between environmental and economic well-being become more apparent.

Of all the technological areas that need to be put on a sustainable basis, energy is the most critical. The generation and use of energy are responsible for a large percentage of almost all forms of pollution. For this reason alone, sustainable development will be impossible without new energy technologies. Energy demands to produce goods and services in industry and transportation and to meet daily needs in residential and commercial buildings have been increasing every year. For example, energy consumption by these sectors in the United States is given in Table 4.1, which shows the relative magnitude and annual changes during 1950–1991. If cheap and nonpolluting energy technologies were in place, a host of new, less polluting and economically attractive industrial, agricultural and transportation technologies could be used.

A phased strategy is necessary to achieve an energy system for sustainable development. The first step in this strategy is greatly increased energy efficiency or conservation. The second

TABLE 4.1   U.S. Energy Consumption, by End-Use Sector, 1950–1991

	Quadrillion Btu			
Year	Residential & commercial	Industrial	Transportation	Total
1950	8.87	15.71	8.49	33.08
1951	9.30	17.13	9.04	35.47
1952	9.54	16.76	9.00	35.30
1953	9.50	17.65	9.12	36.27
1954	9.78	16.58	8.90	35.27
1955	10.41	18.86	9.55	38.82
1956	10.96	19.55	9.86	40.38
1957	10.98	19.60	9.90	40.48
1958	11.64	18.70	10.00	40.35
1959	12.15	19.64	10.35	42.14
1960	13.04	20.16	10.60	43.80
1961	13.44	20.25	10.77	44.46
1962	14.27	21.04	11.23	46.53
1963	14.71	21.95	11.66	48.32
1964	15.23	23.27	12.00	50.50
1965	16.03	24.22	12.43	52.68
1966	17.06	25.50	13.10	55.66
1967	18.10	25.70	13.75	57.57
1968	19.23	26.90	14.86	61.00
1969	20.59	28.10	15.50	64.19
1970	21.71	28.63	16.09	66.43
1971	22.59	28.57	16.72	67.89
1972	23.69	29.86	17.71	71.26
1973	24.14	31.53	18.60	74.28
1974	23.72	30.69	18.12	72.54
1975	23.90	28.40	18.25	70.55
1976	25.02	30.24	19.10	74.36
1977	25.39	31.08	19.82	76.29
1978	26.09	31.39	20.61	78.09
1979	25.81	32.61	20.47	78.90
1980	25.65	30.61	19.69	75.96
1981	25.24	29.24	19.51	73.99
1982	25.63	26.14	19.07	70.85
1983	25.63	25.75	19.13	70.52

TABLE 4.1   U.S. Energy Consumption, by End-Use Sector, 1950–1991 (continued)

Year	Quadrillion Btu			
	Residential & commercial	Industrial	Transportation	Total
1984	26.50	27.73	19.87	74.10
1985	29.73	27.12	20.10	73.95
1986	26.83	26.64	20.76	74.24
1987	27.62	27.87	21.36	76.84
1988	29.00	29.01	22.19	80.20
1989	29.50	29.45	22.38	81.35
1990	28.86	29.90	22.53	81.29
1991	29.56	29.66	22.29	81.51

Source: Annual Energy Review 1991, table 5, p. 15, DOE/EIA-0384(90), U.S. Department of Energy, Energy Information Administration, Washington, D.C., 1992.

step is the use of existing clean fuel technologies. The third and final step is the development and use of advanced, clean, cost-effective energy technologies. The most important examples of such technologies will be discussed in later sections.

Fossil-Fuel Energy

Fossil fuel, the dominant source of energy supply sustaining modern civilization, creates some of the most threatening environmental problems, ranging from global-scale changes in climate to regional effects of acid deposition to local urban air pollution problems and health and safety risks in mining and production. Much of the present technology for energy production will remain with us for at least another fifty years. This means that while working toward technical solutions for sustainability, the most immediate actions relate to clean energy production.

Technologies for the reduction of fossil-fuel combustion emissions are already available and in use, but they are costly, and those emission control technologies for sulfur dioxide, nitrogen oxides and particulates create secondary pollution. A substance of rather recent concern is carbon dioxide, which causes the greenhouse effect, but there is no satisfactory solution as yet. An increase in the energy efficiency of electricity production from fossil fuels, combined with a subsequent shift in fuel from coal to natural gas and later possibly towards biomass, could be the first steps in a strategy aimed at halting the present alarming trends of excessive use of fossil fuels for energy production.

Integrated combined cycle gasification (ICCG) processes offer potential for a substantial increase in electricity production from thermal power plants. Recent technological developments for high-temperature gas turbines have made them an attractive development alternative today. The main advantage of the processes provides a higher operating temperature, which gives a better thermodynamic efficiency for a heat engine. Employing an ICCG process instead of a conventional coal-fired power plant results in much improved energy production efficiency together with reduced emissions.[2]

Nuclear Energy

The biggest question mark in the energy picture over the next few decades is nuclear power. To date, the majority of the general public has opposed it. The difference between fossil fuels and nuclear power is a choice between pollution and the adverse health consequences of fossil-fuel-generated power on one hand and fear of a nuclear accident and consequences of nuclear waste disposal on the other. Two new technological developments could result in making nuclear power safer.

First, within the next decade, a new family of safer nuclear generators is likely to be developed. Second, fusion power may become commercially feasible. Nuclear fusion technology could potentially supply energy with few or no emissions. At the same time, the technology is highly controversial, plagued by problems of reactor safety, waste disposal, weapons proliferation, economic cost and technical reliability. New reactor designs, such as modular high-temperature, gas-cooled reactors, provide ultimately safe reactors; various liquid metal-cooled reactors and advanced pressurized water reactors could lead to so-called passively safe systems and more economical fabrication and installation. These developments, combined with concern over the inevitable emissions of fossil-fuel combustion, could make nuclear power a desirable energy option.

Advanced and Renewable Energy

Advanced energy technologies comprise those based on renewable resources, end-use technologies and hydrogen-based technologies. Renewable resources include a wide range of resources and technologies. Hydro and wind energy sources are among the oldest renewable sources of energy. Photovoltaics is on the cutting edge of scientific and technological research.

Despite the technological advances of the past decade, the cost of renewables in many applications still remains higher than that of fossil-fuel alternatives. Photovoltaic systems, for example, show promise even though they are among the most expensive renewable technology options currently available. Other technologies, however, such as solar heating and wind power, are economically competitive now. The challenge of the 1990s will be to reduce renewable technology costs and to move to full-cost energy pricing so that renewable energy can compete more cost effectively with fossil fuels. But the limited R&D funding for renewables is an obstacle to the development and use of renewable sources of energy.

Strictly speaking, all energy flows are of solar origin, but when we speak of renewable energy production, we generally imply the use of solar radiation for energy production purposes. The sun can be viewed as a blackbody radiator (temperature of

5777 K) with a spectral distribution following Plank's law. The intensity of the solar radiation averages 1353 W/m^2. The atmosphere reduces the intensity of radiation and distorts the spectral distribution somewhat, depending on the optical path traversed by the radiation. The radiation falling on the earth is the ultimate energy source for all activity on earth. The use of solar radiation as a source of thermal energy involves absorption of the incident radiation and conversion of the thermal energy for useful purposes. Energy production involves the use of high-energy photon flux with subsequent transportation of the separated charges in opposite directions for either electricity production (photovoltaics) or chemical reaction (photochemistry).

Direct conversion of solar radiation for energy, although technically feasible, is subject to certain basic limitations. The most serious of these is the dispersion of the energy flow and its variation according to geographical or local conditions. Much of the radiation intensity reaching the earth's surface has been reduced by scattering through the atmosphere and is no longer representative of the photon flux from the regional high-temperature radiation source.[2]

Photovoltaics

Because photovoltaics relies on the virtually limitless and non-polluting solar resource, it offers enormous potential for environmental pollution reduction. There are two basic designs of photovoltaics: single-junction cells and multi-junction cells. The single-junction cell can only be optimized for a small portion of the solar spectrum, determined by the choice of the material. The multi-junction cell consists of several layers of semiconductors, active in different parts of the spectrum, stacked upon each other to yield a better total efficiency in terms of absorbed and converted sunlight. Today, photovoltaics is used in a variety of applications from satellite power generation modules of a few militates to electric utility units as large as ARCO Solar's 6.6-MW power station in California.

Photovoltaic technology has progressed markedly in recent years, resulting in a cost decrease from $15 per kilowatt-hour (kWh) in 1973 to about $0.15 today. Nevertheless, significant technical advances will still be necessary to make photovoltaics competitive with conventional fuel sources. These include new cell designs, such as multi-junction cells that absorb greater portions of the solar spectrum, and new semiconductor materials for improved efficiency and lower cost. Besides these generic developments, manufacturing process improvements, such as micro-fabrication and large-scale applications of thin films, are needed to make photovoltaic cells competitive in mass markets.[1]

Solar Thermal Electricity

Solar thermal-electric equipment converts solar heat to electricity, usually at a central power plant. There are three types of solar thermal technology: trough, central receiver and dish stirling. Trough technology uses reflective troughs to concentrate sunlight; central receiver technology uses tracking mirrors (heliostats) to focus heat from the sun. In both cases, the solar energy is used to produce steam, which then drives a turbine generator. In contrast, dish stirling systems use sun-tracking parabolic dishes to focus heat onto an engine mounted on the dish, which directly converts collected solar heat into electricity.

Most solar thermal electricity generated to date has been from nine parabolic trough plants in southern California. However, central receiver and dish stirling technologies are potentially more cost effective than troughs. Dish stirling technology can be used for a broad range of electricity requirements, especially for needs between 5 kW and 100 MW. Needs less than 5 kW might better be served by a photovoltaic system, whereas needs in excess of 100 MW may be more economically met by a central receiver station.

The environmental benefits of solar thermal electricity are potentially enormous. Like photovoltaics, solar thermal captures a nonpolluting energy source; however, solar thermal probably has narrower applications than photovoltaic technology. Improvement in Stirling engines could make diffusion of solar thermal technology possible for small-scale operations, whereas improvements in heat-transfer fluids would have generic applicability for energy storage in buildings and industry.

In 1990, U.S. solar thermal electric-generating capacity was about 430 MW, primarily from the use of troughs. Solar thermal systems are widely used for water and space heating throughout the world. More than one million homes in the United States have solar-powered water heaters. In the Middle East, rooftop solar collectors provide up to 65% of the energy needed to heat domestic hot water.[1,3]

Biomass

Biomass produced by natural photosynthesis is a good example of solar-driven complex chemical processes. However, photosynthesis is not a very energy-efficient process in nature. Only about half of the incident light falls within the energy domain (0.38 to 0.68 micrometer) that can be used by plants. About 40% of the absorbed radiation energy is lost through reactions (photorespiration) which do not produce carbohydrates. Suppression of such photorespiration reaction is currently a topic of intensive research.

Different species of plants have different efficiencies, and improvement in natural efficiency of photosynthesis appears to be most promising for energy absorption. Biomass as a renewable energy source presents the largest potential for immediate use for producing automobile fuels and vegetable oils. The amounts annually synthesized are roughly equivalent to ten times the world energy demand in calorific terms. The drawback is the same as for most renewable energy sources: it is dispersed and unevenly distributed and is rarely found in great quantities where the energy consumption is largest.

Today, biofuels are a key topic as a possible answer for many severe pollution problems caused by motor vehicles. The most common biofuel is ethanol produced from biomass; the others are vegetable oils and biogas. Ethanol is already produced and used as a fuel in large quantities. Vegetable oils can be produced from the seeds of oil plants such as raps. This oil can be used as a fuel in diesel engines, either directly or after modification. Lubricating oils made from vegetable oils for two-stroke engines seem more attractive in terms of the quantities and costs involved. Biogas is produced as a result of anaerobic fermentation of organic materials such as agricultural residues or urban refuse. Before use as a motor fuel, the methane in the gas must be separated from unwanted constituents such as carbon dioxide, hydrogen sulfide and sulfur-containing compounds. The purified gas can be used directly in Otto engines without modifications or in diesel-type engines with modification.[2]

To make better use of the potential of biomass-based energy production, a thorough systems study of the various opportunities should be done for different regions. Bioconversion is a low-temperature process involving no hazardous chemicals, as opposed to the conventional chemical industry which often employs high temperatures and dangerous chemicals. Further, there is negligible augmentation of the greenhouse effect compared to the burning of fossil fuel, as biomass synthesis is part of the natural carbon cycle. The production of clean automobile fuels is another important area of development for biomass use.

Geothermal

Although naturally occurring hot water and steam formations now provide modest quantities of electricity, hot dry rock, magma and geopressurized technology have considerable potential. World resources of hot dry rock alone are estimated at 100 million quads (quadrillion Btus), which is twenty times all fossil-fuel resources. Because this resource is widely distributed and potentially nonpolluting, its potential risk reduction benefits are very high.

Wind Energy

A typical windmill contains four basic elements: (1) a rotor for capturing the kinetic energy of the moving air, (2) gearbox and transmission, (3) tower structure and (4) generator and control electronics. The most vulnerable part of the windmill is the rotor device, which is sensitive to imbalances caused by rain or snow and ice. Windmills vary greatly in size, from a few kilowatts to more than 4 MW. The average size today is 50 to 100 kW but is expected to rise to 100 to 250 kW with the accumulated experience of the present stations. Technological refinements are producing more efficient windmill generators. An average wind speed of at least 12 miles an hour is needed for these machines to be economical.

Windmills are often used to provide electricity for remote houses, farms or ranches or sometimes small villages. Larger windmills have been built to supply electricity to the grid system. It is also possible to have offshore wind turbine generators providing electricity on land. The cost of producing wind energy is 8 to 12 cents/kWh, depending on the location and size of the unit. The goal of the U.S. program is to lower the cost to 4 cents/kWh. In California, there are more than 15,000 wind turbines producing over 2.5 billion kWh of electricity annually—enough to serve half a million homes.[1,2]

Hydroelectricity

The typical hydropower technology is a dam where the water is stored and regulated. To convert the mechanical energy into electrical energy, the falling water powers a turbine linked to a generator. Hydroelectric plants can have serious environmental drawbacks. The dams needed by the bigger projects flood large tracts of land, upsetting the ecological balance and threatening wildlife. In the tropics, these flooded areas become insect breeding grounds and can spread waterborne diseases, threatening millions of lives.

Among advanced and renewable energy technologies, photovoltaic, wind and solar thermal technologies are demonstrating the potential to contribute a greater share to electricity supply. They could benefit from accurate cost-accounting methods of resource comparison. However, they are intermittent power sources, generally producing electricity only when the wind blows or the sun shines. This limitation makes them less attractive to utility companies, which generally prefer power sources to be dispatchable. Intermittent resources would be more attractive to utilities if cost-effective methods for storing electricity existed. Storage methods such as batteries and compressed air are currently being explored; in the longer term, the use of superconducting magnetic energy storage or hydrogen as an energy storage medium has potential.[3]

Development of wind, photovoltaic and solar thermal technologies confronts several significant barriers: (1) wind and solar resources vary by geographic region and daily weather conditions, (2) they are generally not cost-competitive with fossil fuels under traditional utility accounting and (3) utility companies can be discouraged from developing renewable resources because they perceive their development projects as risky investments.

Energy Storage and Application

Many of the technical options for environmentally benign energy conversion have only limited applicability until energy storage and application are improved. The following technologies represent particularly critical directions for research in this field.[1,2]

Batteries

Improved batteries make possible the use of a wide variety of environmentally superior technologies, such as "zero-emission" vehicles and applications of solar and wind power. Key to the development of improved batteries will be design advances that lead to a reduction in toxic heavy metal components and to the efficient recovery and reuse of these components. Wide-ranging technical options are currently being investigated, including improvements in conventional lead-acid and nickel-cadmium electrochemistries and the development of batteries using aluminum-air, lithium-aluminum-iron-sulfide, sodium-sulfur, sodium-iron-sulfide and other substances. Higher energy density, longer life, lower cost and faster charging are some of the clear technical goals. The improved products and services of new battery technology offer the potential for very high returns on its development. Although considerable research is now underway, the basic challenge is to achieve generic advances in electro-chemistry.

Superconductors

High-temperature superconductors would make it possible to store electricity directly without converting it to chemical, thermal or mechanical forms. If affordable superconducting coils could store electrical current with virtually no resistance loss, the applications would be virtually limitless. Consequently, this technology is already the focus of worldwide R&D efforts to overcome the fundamental technical challenges impeding commercial use.

Superconductors might provide direct environmental benefits by improving energy transmission and indirect benefits by making such intermittent energy sources as solar and wind energy technologies more competitive. On the other hand, the development of the technology may also pose environmental risks if such toxic elements as thallium are used and disposed of inappropriately. Superconductor is a potentially revolutionary generic technology that must be developed and applied in an environmentally beneficial manner.

Heat Storage

Improved heat storage devices and materials could raise the energy efficiency of buildings, solar thermal electric systems and other storage applications. Environmental benefits would result from both decreased energy demand and the availability of new power systems. Although water-based systems have recently become better established, advanced heat storage materials, such as phase-change salts, are far from commercialization. The technical problems involved require both highly generic materials research and the development of specific devices.

Fuel Cells

A fuel cell is a device which allows the conversion of the chemical energy in a fuel directly into electricity, without going through the traditional heat engine cycle. Fuel cells, which are extremely quiet and nonpolluting, can produce electricity at efficiencies greater than that of fuel combustion and could be used in vehicles, houses and industries with considerable environmental benefits. Currently, some applications are commercially viable, but further development and extension beyond limited niches depend on advances in electrochemistry and materials and membrane technology. Phosphoric acid, molten carbonate, solid oxide and proton-exchange membrane cells are among the major types of systems under investigation. None will be viable until costs fall and reliability and performance improve. Both the possibly wide diffusion of fuel cell technology and the generic nature of the technological advances suggest that fuel cells represent a strategic opportunity for the future.

Hydrogen is an entirely nonpolluting fuel whose supply is inexhaustible; it can readily be used as either a liquid or a gas. Hydrogen is produced either by splitting water (requiring electricity) or from other conventional fuels such as natural gas, fossil fuels, or biomass, which creates other emissions. Nevertheless, hydrogen holds great promise for clean and efficient electricity production for mobile use as well as large-scale power generation. Problems with hydrogen storage such as explosive tendencies now limit the feasibility of hydrogen-powered vehicles, which offer potentially high environmental benefits. Although electrolytic and transmission technologies do not appear to present major environmental problems, improvements in metal or organic hydride storage are necessary before hydrogen-powered vehicles become viable. Hydrogen storage illustrates how important generic technical advances can be as a foundation to solve particular environmental problems.[1]

Cost Comparisons of Energy Technologies

Despite the technological advances of the past decade, the cost of renewables in many applications still remains higher than that of fossil-fuel alternatives. Photovoltaic systems, for example, show promise even though they are among the most expensive renewable technology options currently available. Other technologies, however, such as solar heating and wind power, are economically competitive now. The challenge of the 1990s will be to reduce renewable technology costs and to move to full-cost energy pricing so that renewable energy can compete more cost effectively with fossil fuels. But the limited R&D funding for renewables is an obstacle to the development and use of renewable sources of energy.

Utilities typically compare average production costs across their entire grid (rather than at a specific location) when evaluating alternative energy sources. Traditional utility planning and regulatory methods may emphasize the risks of investing in capital-intensive projects while minimizing future fuel price risk. As a result, wind, photovoltaic and solar thermal resources, which have high capital costs and zero fuel costs, may be at a disadvantage compared with fossil-fuel projects, which have lower capital costs but higher and unpredictable costs of fuel plus environmental damages. Table 4.2 shows a comparison of levelized costs of electricity for various energy sources developed by the California Energy Commission.[3]

TABLE 4.2 Comparison of Levelized Costs of Electricity for Various Energy Sources (cents/kWh)

Energy technology	Capital costs	Fuel costs	O&M costs	Total
Coal, 1995	1.2–3.1	2.8–3.3	0.4–0.6	4.5–7.0
Natural gas, 1995	3.2–5.8	3.8–4.3	0.2–0.3	4.9–6.1
Wind, 1995	3.2–5.8	0.0	1.5	4.7–7.2
Photovoltaics, 2000	8.3–16.2	0.0	0.1–0.2	8.4–16.4
Solar Thermal Central Receiver, 2000	3.0–4.6	0.0	2.5	5.6–7.0

Note: (1) Emission control costs of fossil fuel are not included.

       (2) Total costs for each technology are for planning purposes and thus are projected for the year provided as indicated.

Source: GAO, 1993; California Energy Commission, Energy Technology Status Report, Appendix E, June 1991.

INDUSTRIAL TECHNOLOGIES AND POLLUTION PREVENTION

Industrial sustainable technologies can be understood by observing the flow of materials in the course of processing at an industrial site. The most attractive techniques are those that design for the environment through pollution prevention and clean technology. Pollution prevention includes source reduction and environmentally sound recycling. Clean technology avoids toxic materials in preference for less toxic materials when possible; uses fewer raw materials, less energy and water; generates fewer or no wastes (gas, liquid and solid); and recycles wastes as useful materials in a closed system. Clean technology may alter existing manufacturing processes to reduce generation of wastes.

Pollution prevention includes six general categories: improved plant operations, in-process recycling, process modification, materials and product substitutions, materials separations and precision fabrication.[4,5]

Improved Plant Operation

Manufacturers could implement a variety of improved management or "housekeeping" procedures that would aid pollution reduction; they could conduct environmental audits, establish regular preventive maintenance, specify proper material handling procedures, implement employee training and record and report data.

Environmental Audits

Environmental audits may be conducted in many different settings by individuals with varied backgrounds and skills, but each audit tends to contain certain common elements. The practice of environmental auditing also examines critically the operations on a site and, if necessary, identifies areas for improvement to assist the management in meeting requirements. The essential steps include (1) collecting information and facts, (2) evaluating that information and facts, (3) drawing conclusions concerning the status of the programs audited with respect to specific criteria, (4) identifying aspects that need improvement and (5) reporting the conclusions to appropriate management.[6]

The implementation of environmental audits is an especially beneficial activity of initial pollution prevention. Audits enable manufacturers to inventory and trace input chemicals and to identify how much waste is generated through specific processes. Consequently, they can effectively target the areas where waste can be reduced and formulate additional strategies to achieve reductions.

Regular Preventive Maintenance

Regular preventive maintenance involves inspection and maintenance of plant equipment and operational conveyance systems, including lubrication, testing, measuring and replacement of worn or broken parts. Equipment such as seals and gaskets should be replaced periodically to prevent leaks. The benefits of preventive maintenance are increased efficiency and longevity of equipment, fewer shutdowns and slowdowns due to equipment failure and less waste from rejected, off-specification products. Maintenance can directly affect and reduce the likelihood of spills, leaks and fires. An effective maintenance program includes identification of equipment for inspection, periodic inspection, appropriate and timely equipment repairs or replacement and maintenance of inspection records.

Proper Materials Handling and Storage

Proper materials handling and storage ensure that raw materials reach a process without spills, leaks or other types of losses which could result in waste generation. Some basic guidelines for good operational practices are suggested to reduce wastes by:

- Spacing containers to facilitate inspection

- Labeling all containers with material identification, health hazards and emergency first-aid recommendations

- Stacking containers according to manufacturers' instructions to prevent cracking and tearing from improper weight distribution

- Separating different hazardous substances to prevent cross-contamination and facilitate inventory control

- Raising containers off the floor to inhibit corrosion from "sweating" concrete

Spills and leaks are major sources of pollutants in industrial processes and material handling. When material arrives at a facility, it is handled and stored prior to use; material may also be stored during stages of the production process. It is important to prevent spillage, evaporation, leakage from containers or conduits and shelf-life expirations. Standard operating procedures to eliminate and minimize spills and leaks should take place regularly.

Better technology consists of tighter inventory practices, sealless pumps, welded rather than flanged joints, bellows seal valves, floating roofs on storage tanks and rolling covers versus hinged covers on openings. Although these techniques are not novel, they still could lead to large replacement costs if a company has many locations where leakage can occur. Conversely, they could provide large economic benefits by reducing the loss of valuable materials.

Employee Training

Employee training is paramount to the successful implementation of any industrial technical and managerial program. All the plant operations staff should be trained according to the objectives and the elements of the program. Training should address, among other things, spill prevention, response and reporting procedures, good housekeeping practices, materials management practices and proper fueling and storage procedures. Properly trained employees can more effectively prevent spills and leaks as well as reduce emission of pollutants.

Well-informed employees are better able to make valuable waste reduction suggestions. Plant personnel should comprehend fully the costs and liabilities incurred in generating wastes. They should have a basic idea of why and where waste is produced and whether the waste is planned or unplanned.[7]

In-Process Recycling

Materials are processed, frequently in the presence of heat, pressure and/or catalysts, to form products. As materials are reacted, combined, shaped, painted, plated and polished, excess materials not required for subsequent stages become waste, frequently in combination with toxic solvents used to cleanse the excess from the product. The industry disposes of these wastes either by recycling them into productive reuse or by discharging them into the air, water or land. Often costly treatment is required to reduce the toxicity or pollutant impact of the waste discharge before final disposal. These liquid, solid or gaseous wastes at each stage of the production process are the source of pollution problems. On-site recycling of process waste back into the production process will often allow manufacturers to reduce pollution.

Solvents are being recycled in many industrial processes. The current goal of solvent recycling is to recover and refine its purity similar to virgin solvent for reuse in the same process or of sufficient purity to be used in another process application. Recycling activities may be performed either on-site or off-site. On-site recycling activities are (1) direct use or reuse of the waste material in a process, where wastes are allowed to accumulate before reuse and (2) reclamation by recovering secondary materials for a separate end use or by removing impurities so that the waste may be reused.

Off-site commercial recycling services are well suited for small quantity generators of waste that do not have a sufficient volume of waste solvent to justify on-site recycling. Commercial recycling facilities are privately owned companies that offer a variety of services ranging from operating a waste recycling unit on the generator's property to accepting and recycling batches of solvent waste at a central facility.

Process Modification

Many industrial plants have prevented pollution successfully by modifying production processes. Modifications include adopting more advanced technology such as process variable controls, improved cleaning processes, chemical catalysts and segregating and separating wastes, as discussed in the following sections.[8]

Process Variable Controls

Temperature and pressure applications are critical variables as materials are reacted and handled in industrial processes. They can significantly alter the formation of toxins. Improvements include better control mechanisms to meter materials into mixtures; better sensors to measure reactions; more precise methods, such as lasers, to apply heat; and computer assists to automate the activity.

Improved Cleaning Processes

The cleaning of parts, equipment and storage containers is a significant source of contamination. Toxic deposits are common on equipment walls. The use of solvents to remove such contamination creates two problems: disposal of the contaminants and emissions from the cleaning process itself. Some changes include the use of water-based cleansers versus toxic solvents, nonstick liners on equipment walls, nitrogen blankets to inhibit oxidation-induced corrosion and solvent-minimizing techniques such as high-pressure nozzles for water rinse-out.

For example, the Sandia Laboratory has embarked on a program to reduce hazardous liquid waste byproducts of cleaning processes used in the manufacture of electronic assemblies and precious machine parts by (1) alternative solvents used to remove solder flux residues during electronics assembly manufacture, (2) alternative solvents used in ceramic header fabrication and (3) alternative manufacturing processes that eliminate the need for solvent cleaning of precision optical components prior to mounting.[9]

Chemical Catalysts

Catalysts are used throughout the chemical, materials processing and food production industries. The field of catalysis is currently achieving significant advances, with contributions from new developments in chemistry, materials science and biotechnology. In industrial processes and effluent treatment, catalysts can reduce environmental risks by preventing pollution. In process control, catalysts can improve product yield, permit the use of more benign feedstocks or remove undesirable byproducts.

Several research areas appear vital for continued advances in catalysis: better understanding of chemical reaction mechanisms and material surface phenomena, the improved understanding of protein structure and function needed before enzymes can be engineered and the fabrication of highly ordered supramolecular structures which may suggest new catalysts. Since all of these fields are at the basic end of the research spectrum, they are both far from competitive forces and widely applicable to many industrial and environmental problems.

Since catalysts facilitate chemical reactions, they are welcome in pollution prevention research. Better catalysts and better ways to replenish or recycle them would induce more complete reactions and less waste. Substitution of feedstock materials that interact better with existing catalysts can accomplish the same objective.[4]

Materials and Product Substitutions

Materials and product substitutions are complex issues. They include not only technological considerations, but also economic and consumer preferences. Obviously, the use of less toxic materials in production can effectively prevent pollution in a decentralized society. Scientists and engineers are actively evaluating and measuring material toxicity and developing safer materials. Likewise, the life cycle approach requires that products be designed with an awareness of impacts from the raw material stage through final disposal stage. Examples of the product life cycle approach include substitutes for fast-food packaging, DDT, PCBs, CFCs and leaded gasoline.

Materials Substitution

Industrial plants could use fewer hazardous materials and/or more efficient inputs to decrease pollution. Input substitution has been especially successful in material-coating processes, with many companies substituting water-based for solvent-based coatings. Water-based coatings decrease emissions of volatile organic compounds while conserving energy. Substitutes, however, may take a more exotic form, such as oil derived from the seed of a native African plant, *Vernonia galamensis,* to substitute for traditional solvents in alkyd resin paints.[10]

Product Substitution

Manufacturers also could reduce pollution by redesigning or reformulating end products to be less hazardous. For example, chemical products could be produced as pellets instead of powder to decrease the amount of waste dust lost during packaging. Unbleached paper products could replace bleached alternatives. Due to uncertain consumer acceptance, redesigning products could be one of the most challenging avenues for preventing pollution in the industrial sector. Moreover, product redesign may require substantial alterations in production technology and inputs, but refined market research and consumer education strategies, such as product labeling, will encourage consumer support.

Changes in end products could involve reformulation and a redefinition of product requirements to incorporate environmental considerations. For example, the end product could be made from renewable resources, have an energy-efficient manufacturing process, have a long life, and be nontoxic as well as easy to reuse or recycle. In the design of a new product, these environmental considerations could become an integral part of the program of requirements.

Materials Separation

In the chemical process industry, separation processes account for a significant portion of investments and energy consumption. For example, distillation of liquids is the dominant separation process in the chemical industry. Pollution-preventive technology aims to find methods that provide a sharper separation than distillation to reduce the amounts of waste, improve the use of raw materials and yield better energy economy.

Supercritical Extraction

Supercritical extraction is essentially a liquid extraction process employing compressed gases under supercritical conditions instead of solvents. The extraction characteristics are based on the solvent properties of the compressed gases or mixtures. The solvent power of supercritical gases or liquids has been known for more than 100 years, but the first industrial application did not begin until the late 1970s.

From an environmental point of view, the choice of extraction gas is critical, and to date, only the use of carbon dioxide would qualify as an environmentally benign solution. From a chemical engineering point of view, the advantage offered by supercritical extraction is that it combines the positive properties of both gases and liquids (i.e., low viscosity with high density), which results in good transport properties and high solvent capacity. In addit on, under supercritical conditions, solvent characteristics can be varied over a wide range by means of pressure and temperature changes.

Membranes

Membranes are important in modern separation processes for several reasons. They work on continuous flows, are easily automated and can be adapted to work on several physical parameters, such as molecular size, ionic character of compounds, polarity and the hydrophilic/hydrophobic character of components.

Microfiltration, ultrafiltration and reverse osmosis differ mainly in the size of the molecules and particles that can be separated by the membrane. Liquid membrane technology offers a novel method in that separation is affected by the solubility of the component to be separated by a liquid membrane rather than by its permeation through pores, as is the case in conventional membrane processes such as ultrafiltration and reverse osmosis. The component to be separated is extracted from the continuous phase to the surface of the liquid membrane, through which it diffuses into the interior liquid phase. Promising results have been reported for a variety of applications; liquid membrane technology is claimed to offer distinct advantages over alternative methods, but it is not yet widely available.[11]

Ultrafiltration

Ultrafiltration separates two components of different molecular mass. The size of the membrane pores constitutes the sieve mesh, covering a range on the order of 0.002 to 0.05 micron. The permeability of the membrane to the solvent is generally quite high, which may cause an accumulation of the raffinate phase close to the surface of the membrane, resulting in increased filtration resistance (i.e., membrane polarization and back diffusion). However, the application of transmembrane feed flow is being used effectively to reduce membrane polarization.

Reverse Osmosis

Reverse osmosis is generally based on the use of membranes that are permeable only to the solvent component, which in most applications is water. The osmotic pressure due to the concentration gradient between the solutions on both sides of the membrane must be counteracted by an external pressure applied on the side of the concentrate in order to create a solvent flux through the membrane. Desalination of water is one area in which reverse osmosis already is an established technique. The major field for future work will be increasing the membrane flux and lowering the operating pressure currently required in demineralization and desalination by reverse osmosis.

Electrodialysis

Electrodialysis is used to separate ionic components in an electric field in the presence of semi-permeable membranes, which are permeable only to anions or cations. Applications are demineralization and desalination of brackish water or recuperation of ionic components such as hydrofluoric acid.

Adsorption

Adsorption involves intermolecular attraction forces between the molecules in gas or liquid that are weaker than the attractive forces between these molecules and those of a solid surface. Adsorption is the removal of a pollutant from a gas or liquid stream to become attached to a solid surface. For example, gas adsorption processes can be used to separate a wide range of materials from process gas streams. Normally, adsorption processes are considered for use when the pollutant is fairly dilute in the gas stream. The magnitude of adsorption force, which determines the efficiency, depends on the molecular properties of the solid surface and the surrounding conditions.

Adsorbents may be polar or nonpolar; however, polar sorbents will have a high affinity for water vapor and will be ineffective in gas streams that have any appreciable humidity. Gas streams associated with industrial processes will be humid or even saturated with water vapor. Activated carbon, a nonpolar adsorbent, is effective at removing most volatile organic compounds Examples of adsorbents for separating gaseous pollutants include activated carbon, activated alumna, silica gel, molecular sieves, charcoal and zeolite.

Separation Operations

Separation operations present one of the clearest opportunities for preventing pollution by using appropriate technology instead of resorting to remediation after the fact. A number of separation technologies previously discussed are emerging as critical. Better membrane systems, for example, could eventually replace conventional distillation and evaporation processes. Supercritical fluid extraction may obviate the use of organic solvents in many industrial processes and could also help in cleaning up contaminated soil and water. Affinity separation, based on specific binding of particular molecules to a target molecule, could conceivably be used in large-scale efforts to purify dilute products and, in other settings, could be used to remove dilute pollutants.

Precision Fabrication

The ability to manipulate matter precisely by computer controls would allow a great reduction in industrial use of natural resources and industrial emissions. Although precision fabrication of electronic and optical materials is widely recognized as a critical technological capability, its environmental benefits remain largely unnoticed.

A wide diversity of techniques fall under the rubric of precision fabrication. Nanolithography—X-ray, electron beam and ion-beam techniques for etching features on chips—can, with increasing precision, improve chip capacity and quality dramatically. Thin films or precision coating techniques, including chemical and physical vapor deposition and laser and ion-beam implantation, can decrease the cost of photovoltaic cells, improve the performance of electronic components, increase wear and corrosion resistance and make smart building components more widely available. Precision fabrication, like many emerging technologies, draws from various technical fields, including manufacturing, materials and chemical sciences. While many techniques are still precommercial, eventually their use will be widespread.

AGRICULTURAL BIOTECHNOLOGIES

Agriculture is already responsible for widespread and severely negative environmental repercussions such as chemical pollution from pesticides and fertilizers, deforestation, soil erosion and damage to plant and animal life. Biotechnology appears to hold the potential to bring about an environmentally friendly green revolution. Included in virtually all lists of critical technologies, biotechnology not only can reduce environmental risks, but also can be applied generically to a wide range of problems.

Biotechnology is an exceptionally broad technical field. It includes recombinant DNA techniques, protein engineering, monoclonal antibody production and bioprocessing. Much of biotechnology research has been supported with generous public funding. Most funding for biotechnology research has been directed to medical applications. For several reasons, agricultural applications, particularly those with the largest environmental payoff, have been relatively underemphasized. Much of agricultural biotechnology is still highly experimental and precommercial. Thus, the potential reductions in environmental risk that seem possible depend highly on technical advances.

Biotechnology's agricultural applications include gene-transfer techniques, genetic engineering of plants and new approaches to animal breeding and bioprocessing. Many important environmental advances are underway: the development of microbial inoculates that diminish the need for chemical fertilization and pest control, advances in biological pesticides and the modification of genetic characteristics such as nitrogen fixation and photosynthetic efficiency. Some fruits of this research are already at hand or are on the horizon: *Bacillus thuringiensis,* a biorational pesticide that quickly breaks down into harmless components, is well known, and insect-herbicide-resistant corn is expected to be approved and may become a billion-dollar market.

Biotechnology, in general, is vitally important to the solution of nonagricultural environmental hazards. The bioremediation of hazardous wastes, now in limited use, can provide a cost-effective and environmentally superior option. Biosensors, combining information and biotechnology, may become an important element of pollution monitoring. For example, electronic, chemical and biological sensors could all lead to better monitoring of soil, pest infestation levels and animal nutrition. Biotechnology stands as a classic example of the close link between scientific research and technological innovation. Governments must support generic research that private firms cannot appropriate.[1,2]

TRANSPORTATION AND BUILDING ENERGY END USES

We depend on energy to produce goods and services for industry, transportation and residential and commercial buildings. These energy end uses vary significantly in terms of the efficiency improvements that can be made and the incentives and opportunities that energy users have to employ them. Technological improvements in these energy uses offer the greatest strategic opportunities for broad application and substantial reduction in environmental risk.

To ease the transition from today's high energy consumption path to a sustainable path based on renewable resources and other advanced energy technologies, we must make effective use of the cleanest fuels available right now. Natural gas is one choice for such a transition; nuclear power may be another. Current forecasts regarding natural gas fuel prices and availability indicate that supplies will be abundant and economical for years to come. As advanced energy technologies improve power plant efficiencies from the present 30% range to a projected 50 to 60%, the use of natural gas as a transition fuel to generate power makes sense both economically and environmentally.[1]

Approaching the twenty-first century, we must develop even cleaner and more efficient energy technologies, make them cost effective and employ them in every sector of the economy. We must demonstrate the technical feasibility of solar, wind and geothermal energy technologies, introduce new preventive and control techniques and adopt higher energy efficiency standards for everything from automobiles to household appliances.

Energy efficiency is the most cost-effective and environmentally beneficial approach for resolving the nation's major energy problems at this time. Although a totally nonpolluting, nonfossil energy system may be feasible in the future, the most needed and desirable technologies at present are those that increase energy efficiency in the transportation and building sectors.

Transportation

The adverse environmental consequences of transportation are well recognized. Three basic strategies could lessen or eliminate these environmental costs: cleaner vehicles, more efficient vehicle use and decreased travel demand. Radical new technologies that could make a major contribution—for example, electric or hydrogen vehicles—are currently impeded by energy storage problems. However, advanced engine designs (e.g., low heat rejection engines), ceramic engines, improved electronic controls and continuously variable transmissions are among the advances which would be comparatively easy to integrate into the current vehicular fleet.

Transportation accounts for nearly one-third of all U.S. fossil-fuel consumption and produces a large percentage of air pollution. Although vehicle fleet fuel efficiency has improved dramatically in recent years, efficiency gains per vehicle mile are offset by the higher number of vehicles traveling more miles each year. For transportation, the long-range need is a shift in auto technology to less polluting alternative sources of energy. This will require incentives for more fuel-efficient automobiles, advanced energy technologies such as electricity and (in the longer term) hydrogen and for consumers to drive less.

Buildings

Because space heating and cooling, lighting, water heating and appliances account for a large percentage of energy use, making buildings more energy efficient would reduce environmental pollution substantially. Many technological improvements now available—for example, more efficient lighting and appliances—are not being used as broadly and rapidly as is desirable. Current heating and cooling technologies could, in theory, improve efficiency as much as severalfold. Energy-efficient technologies available today can perform the same tasks, produce the same products and provide comfort and convenience comparable to traditional technologies while using a fraction of the energy.

For example, compact fluorescent lamps are "off-the-shelf" products so efficient that an 18-W fluorescent provides as much light as a 75-W incandescent bulb—yet the fluorescent lamp lasts ten times longer and uses only one-fourth as much electricity. Alternative refrigeration cycles and refrigerants, improved controls, capacity modulating systems and thermally activated heat pumps can all offer substantial efficiency gains.

Advanced materials have a myriad of potential applications for high-performance insulation in building shells and other uses. Evacuated powder panels and phase-change salts for thermal storage are two of many examples. In addition, the combination of new materials and information systems could lead to active and smart building or component systems that monitor and adjust to external conditions. Long-term research of energy technologies for buildings may fundamentally alter the concept of building shells and components that could incorporate heat, humidity and light control. These are environmentally important applications of more general building-block technologies.

Individuals as well as industry should be the focus of incentives to encourage environmentally improved technologies. Homeowners should be induced to renovate houses to make them environmentally sound. Automobile purchasers should be induced to consider more fuel-efficient vehicles. There is virtually no area of the economy that could not be motivated to renovate, update or replace environmentally harmful technologies via the right incentive.

DESIGN FOR THE ENVIRONMENT

Design for the environment is a new concept and challenging frontier for engineering professionals. Environmentally compatible products minimize the adverse effects on the environment resulting from their manufacture, use and disposal. Environmental considerations during product planning, design and development can help industries minimize the negative impact of the products on the environment. While changing product design to prevent pollution, engineering professionals should maintain the quality or function of the product. Design for the environment can be achieved by the people directly involved within the framework of company policy and with support from company management. It can be implemented through (1) product life cycle assessment, (2) product life cycle design and (3) product life cycle costing analysis, as shown in Figure 4.2.[12]

Increasingly, businesses and institutions are assessing their environmental performance comprehensively through tools such as environmental auditing. They recognize that good environment performance does not come simply from attempting to improve in one or two areas. The development of a product

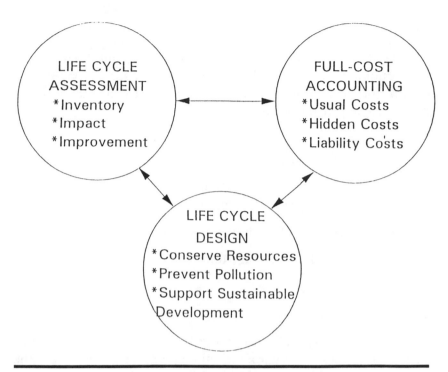

FIGURE 4.2 Designing environmentally compatible products.

which is environmentally safer in use or disposal may well be desirable, but if its production consumes four times as much energy, the net result may be less beneficial.

It is difficult to be aware of all the possible environmental issues which surround design decisions. In order to keep pace with rapid developments in scientific understanding and the discovery of new causes for concern, the designer must stay in touch with the environmental agenda, by following media coverage, joining environmental professional organizations and participating in meetings and seminars/workshops provided by professional or trade associations. It is more difficult still to find detailed information on the environmental performance of alternative materials or processes or to identify sources of supply which are guaranteed to be environmentally conscious.[13]

Plant Design

Incorporating environmental constraints into all plant design stages will require a new generation of design procedures, formal education and continuous training. The essential elements of these new design procedures are:

- Procedures for selecting environmentally compatible materials

- Design of unit operations that use less energy, fewer resources and minimize waste

- Economics of pollution prevention

- Process flowsheeting for minimizing wastes[14]

Product Design

In both the redesign of existing products and the design of new products, additional environmental requirements will affect the methods applied and procedures followed. These new environmental criteria will be added to the list of traditional criteria. Environmental criteria for product design include:

- Increasing energy efficiency and reducing energy consumption

- Using renewable natural resource materials

- Using recycled materials

- Using fewer toxic solvents or replacing toxic solvents with less toxic solvents

- Reusing scrap and excess material

- Reducing packaging requirements

- Producing more replaceable component parts

- Producing more durable products

- Producing goods and packaging that can be reused by consumers

- Manufacturing recyclable final products

Materials Design and Processing

The current revolution in materials design and processing involves not only the development of new and superior metals, polymers, ceramics and cosmetics, but also radically new ways of producing them. The trajectory of development is wide ranging, merging materials sciences with chemistry and precision manufacturing with biotechnology. Given these trends, critical technologies for design include materials synthesis and processing and the development of electronic and photogenic materials, ceramics, composites and high-performance metals and alloys. The materials revolution has myriad implications for the environment. New methods of materials processing, such as direct steelmaking, electrochemical processing of metals and chlorine and microbial mineral processing, could dramatically diminish pollution and energy requirements.

Systems Approach

Preventive and clean technologies are necessary as part of a systematic approach that focuses not just on the parts, but also on the internal consistency of the system as a whole. Preventive technology includes process waste minimization, recycling and reuse, whereas clean technology emphasizes input materials and energy and also output products that are environmentally acceptable. The systems approach recognizes the many elements of preventive and clean technologies as well as economic competitiveness and the need to incorporate both existing and emerging technologies.

In fact, sustainable development technology is a combination of efficient product designs, operational procedures that use materials with minimal waste and toxic byproducts, and technical processes that are precisely monitored and controlled. Much of the technology to prevent pollution is available now and is frequently not complex. Among current techniques, much headway has been made in the area of solvent control and coating/painting.

The Social Context of Design

Environmental protection is not simply a practical issue; it is also a moral one. As part of the professional code of practice, designers should aim to minimize the environmental impact of their work. Environmental ethics have been accepted in most professional societies and several institutions. Designers have been urged to take responsibility for a green ethic because of the central role they play in influencing the environmental performance of so many things. Designers can make a broad environmental contribution because they are able to identify and solve problems through their insight and mastery of a wide range of valuable skills. Designers have always striven for a better way of doing things—now that better way can embrace environmental, ethical and social issues, too.

Very many other moral issues are now emerging as possible influences on public purchasing and behavior. Advocacy groups provide consumers with more information about the activities of companies and governments, and companies are beginning to use their corporate behavior to establish a desirable image and achieve further differentiation in the marketplace. Hence, the welfare of animals, fair trade with Third World countries, the treatment of minorities, military expenditures and many other issues are rapidly becoming part of the marketing agenda.

It is beyond the scope of this chapter to begin to examine the role of the designer in addressing this wider sphere of ethical and social issues. But it is not unrealistic to expect that, irrespective of the designer's own moral code, ethical issues will present additional criteria in the design process through customer behavior and preference. Once again, the central, influential position of the designer provides a real opportunity to effect change.

Training and Education for Design

Many design courses continue to virtually ignore environmental issues and thus equip students poorly even for current demands. There can be little doubt that it is essential to include information about environmental issues and their relationships to the design process in the core curriculum. Equally important is the development of an understanding that the environment is not a separate subject in which designers can choose to be interested or not, but rather is one basic criterion against which all design work should be assessed.

The relative newness of design for the environment means that good teaching material is only just emerging, and there is a lack of good case histories to form the basis of projects. This may mean that students will have a real opportunity to conduct original research and to produce material which may be of practical use beyond its value to them as a learning experience. The redesigning of our society to minimize environmental problems will demand an integrated approach—one which recognizes that everything is interrelated. Breadth of experience and vision will be valued as we attempt to cope with large-scale and highly complex problems.

THE ROLE OF DESIGNERS

Environmental problems, or risks to health and safety, are often hard to foresee. It will, however, become an increasingly important aspect of the designer's work to minimize the risks arising from the failure of a product or process. Much of the focus must be on the design of not only products and services, but also industrial facilities or building developments. Environmental impact assessments must anticipate the consequences of modifications to the original design or of human negligence. The eventual environmental impact of a major construction, such as an industrial processing plant or industrial park development, must be carefully assessed at the planning stage to minimize the risk of long-term damage to the local ecology. Sophisticated electronic monitoring instruments and sensing devices may be used increasingly to detect the presence of undesirable environmental pollutants and to prevent accidents.

Designers must consider a wide range of criteria, such as design process marketing, production, financial and technical considerations. Compared with these, environmental considerations could be even more complex and hard to handle. There are very often no clear answers because information is hard to find; guidelines may not be available. Much original research and thinking may be necessary. Designers cannot be expected to have the time or specialized knowledge to gather all the information they will need and, indeed, some information may take several years of research to emerge.

The designer must therefore be responsible for asking the right questions and raising the relevant issues. By demonstrating an awareness of potential problems, the designer is helping minimize risks to the success of the project or program. A review of environmental auditing or assessment data may become as essential for the background briefing of the designer as knowledge of the company's marketing strategy or institutional philosophy and objectives.

Designers have never been able to work in isolation; they are necessarily team workers, if sometimes reluctantly. The need to consider environmental issues and execute details at an early stage in the design process will mean that close partnerships with other professionals in the project or program will be essential. On environmental issues, the designer must be in a position to assess and judge the recommendations and output from a wide range of technical specialists.

Designers may find themselves at the forefront of identifying problems which must be addressed by technology. Sometimes existing technologies may not be able to provide the solution, and designers may have to stimulate and influence the development of a new technological approach. The role of the designer as the stimulus for, rather than the distant incorporator of, technological advances will again prove challenging and will require interests and skills which may go beyond those areas traditionally regarded as appropriate in design. Just as designers must monitor the environmental agents carefully, so they must also follow technological developments to be sure of incorporating the most environmentally advanced technology or to identify gaps which need to be researched.

If they choose to do so, designers have an opportunity to exert considerable influence. But this influence will have to be supported by knowledge, open-mindedness and flexibility, and an ability to go on learning. Designers can now prove that they are essential partners in planning and implementing a sustainable future.

THE ROLE OF GOVERNMENT

The role of the government in developing and adopting technology for sustainable development will be crucial. There is an opportunity to implement policies to stimulate investment in new environmentally beneficial technologies and to encourage phasing out of outmoded capital equipment. Three essential recommendations worthy of consideration relate to regulations affecting technology, incentives and government–private sector cooperation.

Regulations affecting technology should be based on extensive and unbiased assessments of the technological possibilities,

associated costs and effectiveness in promoting technology development and use. Regulatory requirements can produce innovative solutions to environmental problems, such as the labeling
of major appliances for energy efficiency. Product labeling can
increase public understanding and may direct market forces towards more environmentally benign technologies and products.

Government agencies must provide clear criteria and specific
incentives, such as new taxes, carefully crafted investment tax
credits or "green" research and development tax incentives, to
stimulate savings and investment in environmentally advanta-
geous technologies. Regulatory obstacles to the development and
introduction of environmentally superior technologies should be
removed.

Government agencies must cooperate and collaborate with
the private sector to inspire positive technological change. This
is particularly true in areas where the benefits of investing in
research and development cannot be directly recouped by individual companies. Together with business, the national government must also join international sustainable programs to facilitate worldwide adoption of new technologies at lower cost for
a given country. A separate international effort for sustainable
technology development could also help less developed countries bypass the heavily polluting technologies that have characterized development over the past century in industrialized nations and direct their resources toward using the developing
sustainable technologies.

INFORMATION AND COMMUNICATIONS

Development of information technologies has probably been
more rapid and dramatic than progress in any other field within
the last thirty years. This dynamic progress may continue into
the foreseeable future. Information technologies are often seen
as the new bedrock on which modern societies depend. From

an environmental point of view, the benefits associated with increased applications of information technology are numerous. Improving environmental quality depends increasingly on developing and deploying information technologies, such as computerized monitoring and controlling processes and waste generation. It is obvious that understanding and managing complex global environmental problems, such as climate change and greenhouse effects, require vast quantities of observational information. Such information must be recorded, stored, analyzed, compiled, manipulated and displayed in ways convenient to scientists, policymakers and the public. These requirements cannot be met without significant advances in information technologies.

End-users' demands for information about products and services will increase steadily as people begin to make choices on the basis of more complex criteria. Information about the environmental impact of a product (which may have to encompass ingredients, manufacturing method, use and disposal instructions) will be included on packaging or in accompanying leaflets. Information about the product or service will become a part of the product offer that is valued in its own right. It may therefore assume a much higher profile than in the past.

The development of information systems which are at once comprehensive, easy to absorb and motivating will provide major advantages for product marketing and consumer choices. The growth of products which are sold in many different countries will pass the challenge of how complex information can be communicated, for example, by the use of visual symbols. Official ecolabeling systems will be one aspect of information provision, but these will not cover all categories and will necessarily be a summarized version of more detailed information which may well have to be communicated in some other form elsewhere.

An increasing use of computers and of computer-dependent services such as environmental services will create a demand for designers who can improve the user-friendliness of both hardware and software. The design of on-screen information can be significantly improved by the application of principles based on an understanding of how people react to visual stimuli. "Intelligent" products and new services based on information will meet with wide success only if information can be provided in a highly accessible, sympathetic and attractive way.

CONCLUSIONS

Technology can cause environmental problems; technology can also provide environmental solutions. Development of technology in the past has led to deterioration of the natural environment and the quality of human life. If environmental pollution prevention is to prevail over environmental control and remediation, the future must portend and deploy a wide array of new and more efficient technologies that minimize stress on the environment.

As we approach the twenty-first century, we must develop more efficient and even cleaner technologies, make them cost effective and employ them in every sector of the economy. Technologies for sustainable development are critically needed in energy, industry, agriculture, transportation and land use sectors. The most important and immediate target is to encourage the development and adoption of technologies compatible with pollution prevention and clean technologies. Design for the environment must take its place alongside cost, safety and health as a guiding criterion for technology development. It is imperative that government agencies provide incentives for the development and adoption of sustainable technologies and also participate in partnerships with private industry to explore new and innovative clean technologies. Government agencies should also launch a coordinated international research effort to develop environmentally sound technologies.

Breaking through the barriers to developing and using new technologies requires government leadership, private sector ingenuity and public support. This will necessitate new pricing mechanisms, investment incentives, regulatory changes and collaborative undertakings. Both industry and the government must assess the effectiveness of new technologies in meeting the goals of pollution prevention, waste reduction and economic efficiency to ensure that any new technology is an environmentally sound technology. Government agencies should increase substantially funding for research and development in sustainable technologies.

Sustainable development requires sustainable technologies. The benefits of these measures notwithstanding, economic signals remain the most important factor driving the development and adoption of environmentally compatible technologies. Thus, it is essential that economic signals reflect environmental values.

ACKNOWLEDGMENTS

The author acknowledges Dr. Marian Mudar of New York State Department of Environmental Conservation for her editorial task and creative comments on the chapter.

REFERENCES

1. NCE, *Choosing a Sustainable Future. The Report of the National Commission on the Environment,* Island Press, Washington, D.C., 1993.

2. A. Johansson, *Clean Technology,* Lewis Publishers, Ann Arbor, MI, 1992, Chapter 4.

3. GAO, Electricity Supply—Efforts Under Way to Develop Solar and Wind Energy. A Report of U.S. Government Accounting Office, GAO/RECD-93-118, April 1993, pp. 14–56.

4. OTA, Serious Reduction. Report published by Congressional Office of Technology Assessment, Washington, D.C., 1987, pp. 78–82.

5. T.T. Shen, *Industrial Pollution Prevention,* Springer-Verlag, Heidelberg, Germany, 1994, Chapters 4, 5, 10, 12 and 15.

6. ICC, Environmental Auditing, International Chamber of Commerce, ICC Publication 468, ICC Publishing S.A., Paris, 1989.

7. G. Hunt et al., Accomplishments of North Carolina Industries—Case Summaries, North Carolina Department of Natural Resources and Community Development, Raleigh, NC, 1986, p. 24.

8. USEPA, "Designing environmentally compatible products of facility," in Pollution Prevention Guide, EPA/600/R-92/088, Office of R&D, Washington D.C., May 1992, Chapter 7.

9. M.C. Oborny et al., Sandia's Search for Environmental Sound Cleaning, Sandia Laboratory Report, Sandia, NM, 1990.

10. M.S. Reisch, "Demand puts paint sales at record levels," *Chem. Eng. News,* p. 41, Oct. 30, 1989.

11. M. Saari, Prosessiteollosuuden Erotusmenetelmat, VTT Research Note, Technical Centre of Finland, Helsinki, 1987, p. 730.

12. T.T. Shen, "Designing environmentally compatible products," in Proceedings of the 7th Pollution Prevention Conference, Albany, NY, June 1–3, 1994.

13. D. Mackenzie, *Design for the Environment,* Rizzoli International Publications, New York, 1991.

14. T.D. Allen et al., *Pollution Prevention: Homework & Design Problems for Engineering Curricula.* A manual published by the Department of Chemical Engineering, University of California at Los Angeles, 1992.

5

HOW DO WE KNOW WHAT IS SUSTAINABLE? A RETROSPECTIVE AND PROSPECTIVE VIEW

John L. Warren

Taking the concept of sustainable development and making decisions based upon it is quite difficult and is something few of us have experience with. This chapter provides background on two related issues: what types of indicators we can use to assess progress towards sustainability from particular actions and what questions we can ask as we design new programs to ensure that they reflect our concerns about sustainability. One is retrospective and the other is prospective.

BACKGROUND

There is a broad range of responses as to what the definition of sustainable development means. This is primarily due to the

John L. Warren is a Senior Program Manager at the Pacific Northwest Laboratory, operated by Battelle for the U.S. Department of Energy. His research is in industrial ecology, the development of eco-industrial parks and the practical application of sustainable development principles to corporation decision making.

©St. Lucie Press CCC 1-57444-079-9 1/97/$100/$.50

values of the organizations and individuals involved in environmental management and global sustainability. In spite of this seemingly disparate range of views, there are several themes and issues that emerge from a review of the sustainable development literature. These are important to consider as we address the type of indicators and questions to use in operationalizing the sustainable development concept.

- *The economy and the environment are inextricably linked.* No human activity can be conducted without some connection to the environment, whether the provision of clean water and food or the latest version of a personal computer.

- *We need to take the "long view."* The sustainable development framework or policy concept entails thinking far into the future and how our present actions might affect our ability to live a wholesome and fulfilling life. This raises the issue of intergenerational equity.

- *Think in terms of systems, feedback loops and boundaries.* This can include ecosystems (both industrial and natural), total energy systems, social systems and economic systems. Industrial ecosystems include all those systems necessary for supporting industrial output: natural, economic, social and technological systems.

- *The scale of human activities is important both spatially and temporally.* We are interested in how activities impact not just our own backyard but also the "global backyard." In addition, activities may have impacts that span decades and beyond or have impacts that are not felt until future generations. One can look at sustainable development on a global, regional, national, state, local and even individual scale.

- *Human actions must be grounded in an enhanced understanding of how the world works and how we can work with the world.* Ecological and ecosystem concepts should be an integral part of decision making.

- *Sustainable development occurs within a dynamic and evolving set of interlocking systems: ecosystems and economic and social systems.* These systems co-evolve to fill niches, thus changing dynamically over time. The world and human response to it (typically which are evolving technologies and belief systems) are not in a static equilibrium. Change is the norm, not the exception!

- *Thinking about sustainable development requires an interdisciplinary approach to addressing environmental and human problems on the earth.* No one discipline can provide the particular perspectives, experience, expertise and tools to address the wide range of challenges in moving toward a sustainable future.

In order to operationalize the concept of sustainable development, organizations and the individuals within them need to address many issues related to what is sustainable and how one knows if an institution is operating in a sustainable manner. This chapter addresses two ways of doing this: the retrospective use of sustainable development indicators and the development of a set of sustainable development questions to use prospectively when assessing proposed or existing activities within the institution.

Assessing the sustainable characteristics of an activity is often difficult because there will be a transition from the existing less than sustainable systems to a more sustainable set of systems and activities in the future. Thus, an organization may be improving the efficiency of systems based on nonrenewables; yet we need to do that in order to provide for an orderly transition to a renewables-based future. Assessment during the transition will be different from after "we have arrived." Furthermore, this transition might actually have very positive outcomes that cannot be measured until future generations.

Harvey Brooks observes:

> ...consumption of nonrenewable, natural resources can actually improve social and physical infrastructures and increase the expansion of intellectual capital (e.g., scientific knowledge and engineering know-how). This, in turn, could make future generations materially richer and more capable of addressing their own environmental challenges.[1]

> Sustainable strategies do not necessarily always entail sacrificing the present for the future. To the extent that we can use a given quantity of presently known natural resources to create more wealth and more useful knowledge with which to find, extract, and use future resources more efficiently, or to substitute more abundant raw materials for those presently used in human artifacts, we may have not only not sacrificed the present for the future, but we may have actually enhanced future options beyond what would have existed in the absence of this draft on present resources....Knowledge, especially knowledge that becomes fully internalized in the skills and capacities of many people widely spread around the world, and is passed on to the next generation, is a resource which is not depleted by more intensive utilization for human benefit today.[2]

INDICATORS OF SUSTAINABLE DEVELOPMENT

Because of these unique characteristics of time and spatial scales, sustainable development metrics must be developed with care. We may have numbers, but they do not tell us what we need to know. The illusion of certainty is more dangerous than the certainty of ignorance. Nonetheless, a variety of groups are developing measurements reflective of sustainability. Some have addressed what makes a good indicator. The following list of criteria is important to consider when developing meaningful indicators particular to a specific set of activities and further fine-tuning for institution-specific needs.

Criteria for Good Sustainable Development Indicators

- Reflect something basic and fundamental to the long-term economic, social or environmental health of a community over generations

- Recognizable and clear: simple, can be understood and accepted by the community

- Quantifiable

- Sensitive to change across space or within groups

- Predictive or anticipatory

- Reference or threshold values available

- Reveal whether changes are reversible and controllable

- Relatively easy to collect and use

- Quality aspects: methodologies used to develop an indicator must be clearly defined, accurately described, socially and scientifically acceptable and easy to reproduce

- Sensitivity to time: if applied each year, the indicator can show representative trends[3-5]

Several groups have developed indicators directly or indirectly related to assessing sustainable development progress. Within the U.S. government, this effort has centered in the Interagency

Working Group on Sustainable Development Indicators. Departments represented include Energy, Environmental Protection, Agriculture, Commerce, Interior, State, Education, and Transportation. An interim report reviewed sustainable development indicators, development of an indicator framework and future plans for the group.[6] This report provides an excellent overview of federal government sustainable development indicator development activities.

Some examples of the development of indicators include:

Bureau of the Census, Bureau of Economic Analysis—In April 1994 released its Integrated Economic and Environmental Satellite Accounts (IEESAs) in which natural and environmental resources are treated like productive assets.[7] Hailed in the press as the "Green GDP."

President's Council on Sustainable Development—Considering best role for the council to play.

EPA—Developing national environmental goals for Earth Day 1995.

Interagency Committee on Environmental Trends—Focuses on product-oriented data collection and exchange. State and local groups include the City of Seattle, which has developed a set of forty indicators specific to Seattle,[3] and the State of Vermont, City of Jacksonville, Florida, and Ventura County, California, which have developed local sustainable development indicators.

World Resources Institute—Publishes data and indicators in biennial report on world environmental resources.

Worldwatch Institute—Publishes annual *State of the World and Vital Signs* with indicators related to environment, resources, economics and social issues.

Canadian "Green Plan"—Called for national environmental indicators to be developed to measure Canada's progress in achieving its environmental goals.

Netherlands—Developing sustainable development indicators as part of the Netherlands Environmental Policy Plan (NEPP).[5]

United Kingdom—Will establish a working group to produce indicators by 1996.

Additional international groups working on environmental and sustainable development indicators include the U.N. Commission on Sustainable Development, U.N. System of Environmental and Economic Accounting (SEEA), UNSTAT (Statistical Office of the United Nations), United Nations Environmental Programme (UNEP), Scientific Committee on Problems in the Environment (SCOPE), Organization for Economic Cooperation and Development (OECD) and The World Bank.

Obviously, many groups are wrestling with sustainable development indicators. Indicators need to be tailored to the scope and scale of what an institution wants to accomplish within the context of its mission. Critical questions to consider when developing means of measuring sustainable development progress include:

- Why do we need the indicators?

- Who will use them and for what purpose?

- How often do we need to update them?

- What does "successful sustainable development" mean for our institution?

- What are appropriate indicators or quality metrics?

- Will some indicators require input from other institutions or regulatory agencies?

- How do we establish a metric for something that occurs in the next generation?

- Are scientific indicators inherently different from those used to track program effectiveness?

- What is important that might not be able to be captured in a seemingly quantitative metric or indicator?

Global Indicators of Sustainability

The following are examples of indicators of sustainable development developed for three spatial levels: global (Worldwatch Institute Vital Signs), national (Netherlands National Environmental Policy Plan), and local (Sustainable Seattle). They can serve as a guide in developing indicators for your organization. In its annual Vital Signs report, the Worldwatch Institute has selected the following key indicators:

Food Trends
- Grain production
- Grain stocks
- Soybean harvests
- Grain used for feed
- Meat production
- Fish catch

Agricultural Resource Trends
- Grain area
- Fertilizer use
- Irrigation

Energy Trends
- Oil production
- Wind power
- Nuclear power
- Solar cell production
- Natural gas
- Energy efficiency
- Geothermal power
- Hydroelectric power
- Coal use

- Carbon efficiency
- Compact fluorescent lamps

Atmospheric Trends
- Carbon emissions
- Global temperature
- CFC production

Economic Trends
- Global economy
- Third World debt
- International trade
- Steel production
- Paper production
- Advertising expenditures

Transportation Trends
- Bicycle production
- Automobile production
- Air travel

Social Trends
- Population growth
- Cigarette smoking

- Infant mortality
- Child mortality
- Refugees

Military Trends
- Military expenditures
- Nuclear arsenal

National Indicators of Sustainability

On a national basis and as part of its National Environmental Policy Plan, the Netherlands has developed measures of environmental policy indicators that are similar to those that might be used for sustainable development assessments:[5]

Change of Climate
- CO_2 emissions
- CH_4 emissions
- N_2O emissions
- CFC production and use

Depletion of Ozone Layer
- Production of CFCs
- Production of halons

Acidification of the Environment
- SO_2 deposition
- NO_x deposition
- NH_3 deposition

Eutrophication of the Environment
- Phosphorus
- Nitrogen

Dispersion of Toxic Substances
- Agricultural pesticides
- Other pesticides
- Priority pollutants
- Radioactive substances

Disposal of Solid Waste
- Quantity of solid waste

Disturbance of Local Environments
- Percentage of Dutch people adversely affected by odor or noise

Local Indicators of Sustainability

Finally, at the local level, a grass-roots group, Sustainable Seattle, has been working to develop a meaningful set of metrics for assessing how well their community is doing. From an initial list of one hundred potential indicators developed through extensive community participation, they selected forty indicators for which to develop data. Twenty of these have been utilized with data; they include:[3]

Environment

- Wild salmon runs through local streams
- Number of good air quality days per year
- Percentage of Seattle streets meeting "pedestrian-friendly" criteria

Population and Resources

- Total population of King County
- Gallons of water consumed per capita in King County
- Tons of solid waste generated and recycled per capita per year in King County
- Vehicle miles traveled per capita
- Gasoline consumption per capita
- Renewable and nonrenewable energy consumption per capita (in Btus)

Economy

- Percentage of employment concentrated in the top ten employers
- Hours of paid work at average wage required to support basic needs
- Percentage of children living in poverty
- Housing affordability for median and low-income households
- Per capita health expenditures

Culture and Society

- Percentage of infants born with low birth weight

- Juvenile crime rate
- Percent of youth participating in community service
- Percent of population voting in primary elections
- Adult literacy rate
- Library and community center usage rates
- Participation in the arts

PROSPECTIVE SUSTAINABLE DEVELOPMENT QUESTIONS

Indicators can be excellent guides to both the present state and the effects of past activities. However, indicators do not serve as a means of improving future actions to make them more "sustainable." As part of a strategic planning effort for the U.S. Department of Energy, a team developed a preliminary set of questions to use in assessing future actions for sustainability. The intent was to enable a user to attain an appropriate level of sustainability based upon his/her values and those of the organization. Consequently, there are no right answers; rather, users can explore the range of options.

The following are "prospective sustainable development questions." Note that the designation "P3" stands for "policy, program or project."

Q1. Sources, flows and sinks of energy and related materials

- What are the energy sources and sinks for this P3 (i.e., where is the energy coming from and where is it going)?

- What is the type of energy sources?

- Are we using stock or flow energy resources?

- How are we using energy in the P3?

- Can you draw a flowchart of the significant flows of energy, materials and information for the P3? If not, how do you know what might happen?

Q2. Rates of change

- Is the rate of creation of the energy resource in balance with the rate of destruction for the P3? Most natural systems exhibit some balance between creation and destruction. Rates of energy creation and dissipation are important elements for sustainability issues.

Q3. Propensity for recyclability

- Does the P3 system infrastructure accept recycled material inputs and deliver recyclable outputs?

Q4. Response time of systems

- How do different systems respond to one another?

- Over what time horizon will impacts occur?

- What is the relevant time frame for the decision?

Q5. Geographic scale of impact

- At what geographic scale will the P3 have impact? An impact at any scale will affect other scales, both larger and smaller.

Q6. Life cycle analysis

- Does the P3 account for the full life cycle impacts and costs of both actions and inactions over broad temporal and spatial scales?

Q7. Values

- Each of us has a particular way or ways we look at the world based on our culture, education, experience and values. Does your particular perspective affect how you view sustainable development for this P3?

Q8. Use of renewables

- What renewables and nonrenewables will be used?

- What particular characteristics are they being used for?

- Could a renewable or more resource-efficient technology provide the same service?

- Does the P3 promote self-sufficiency by using local and renewable resources?

Q9. Equity

- Does the P3 appear to support human life and welfare?

- Does it show a net benefit to society? Do some benefit more than others?

- Are burdens borne by the people who receive the benefits?

- Does this benefit change over time? Present? Future? Scale?

- Is the P3 deployment fair?

- Is there an equitable distribution of costs and benefits? To what extent? Locally? State? Regionally? Nationally? Globally? Present? Future?

Q10. Decision making

- Does the P3 create incentives or disincentives to encourage individual decision makers to incorporate environmental effects into their energy decisions?

- Will the P3 encourage each individual to make sustainable energy decisions or does it merely dictate a change in behavior?

Q11. Consequences

- For a range of human and environmental spatial dimensions—community, county, watershed, country, global environment—what are the effects of the P3?

- Which level is most likely to experience any gains?

- Which level is most likely to suffer negative impacts?

- Are the negative consequences of the P3 reversible? If so, over what temporal and spatial scale?

Q12. Diversity

- Does the P3 contribute to diversity, either cultural or biological? System?

Q13. Social impacts

- Does the P3 support human culture, including religious, ethnic and other values?

- How are social systems affected by the P3?

Q14. Mitigation

- Does the P3 incorporate mitigation strategies for affected entities and people?

Q15. Economic impacts

- If environmental damage is expected from the P3, what are the economic impacts?

- If economic changes occur, what new incentives for changed environmental behavior are created?

- Will markets be affected? If so, how?

Q16. Eightfold questions

The following eight questions (developed by James Wise at Pacific Northwest Laboratory) are similar to the "Eightfold Way" of particle physics.

16.1 What negative feedback (or self-control) is built into the operation of the P3 (e.g., does what is undertaken incorporate the observance of limiting conditions and the means to stay within them)?

16.2 Does this P3 promote a "stable equilibrium" of the elements it affects, or does it drive any of them to an unsustainable growth rate in order to satisfy a derived, consequential need?

16.3 Is this P3 undertaken with the understanding that its immediate products are temporary and, thus, secondary, and that its full implications need to be seen as something qualitatively different?

16.4 Does this P3 work with naturally occurring forces within the system where it will be applied?

16.5 Does the P3 serve more than one purpose or use, so that it efficiently achieves multiple goals?

16.6 Does the P3 minimize the production of waste products at all levels where it operates and encourage the reuse and recycling of base materials and subsequent products?

16.7 Does the P3 encourage symbiotic relationships locally?

16.8 Is the P3 basically compatible with the biology of humans and nature, or will it in any way result in unhealthful impacts to the life and functions of people, plants and animals?

Q17. Harmonization of human laws with natural law

- Human laws of economics and jurisprudence often are not in harmony with the planetary laws based on physics, chemistry, biology, etc. Nothing in the natural world operates to maximize present value to the exclusion of all else. Does a particular decision account for the full life cycle impacts and costs of both actions and inactions over broad temporal and spatial scales?

Q18. Security

- Does the P3 promote a sense of security among beneficiaries? Among others?

Q19. Institutional learning

- Does the P3 provide for learning from what is done and passing learning on to others?

Q20. Primary productivity

- Can you estimate how this P3 would impact human use of terrestrial net primary productivity?

Q21. Prioritization

- Is the framework (ecosystem and human) into which the P3 must fit well defined?

- If there are multiple objectives, are they prioritized?

Q22. Efficiency

- Is it efficient, thermodynamically and economically?

- Is it feasible within real-world institutional constraints and opportunities?

- Does the P3 encourage an energy-efficient technology that may result in energy-inefficient behavior?

Q23. Information

- What information is being transferred to improve decision making and change attitudes toward resource use, efficiency and sustainable development?

Q24. Systems Involved

- What systems are involved: social, environmental, economic, government, other?

- Are we overlooking one that might impact the sustainability of an activity?

SUMMARY

As organizations address what sustainable development means for their present and future activities, they need to recognize the importance of both retrospective and prospective views. Assessing the extent to which past activities have been sustainable or conducted in a sustainable manner requires the use of indicators or other measures of effects that are thought to be associated with sustainable actions. Planned actions need to be viewed through a sustainable development "lens" to determine whether they may or may not be sustainable. Because natural and human systems are inherently dynamic and complex adaptive systems, it is difficult to forecast precisely what may happen.

Consequently, we have proposed the use of a series of probing questions that can serve as a basis for thinking about the sustainable aspects of any program, policy or project. Though these questions have no "correct" answers, they provide a basis for enhancing the overall level of sustainable development.

ACKNOWLEDGMENT

The author is responsible for any errors or omissions. However, many of the ideas in this chapter were developed by a team working with the U.S. Department of Energy's Office of Energy Efficiency and Renewables. The team was part of an effort to provide support for the office's strategic planning initiatives. Team members included Kathy Kreiter, Washington State Energy Office; Dave Bassett, U.S. Department of Energy; Meir Carasso, National Renewable Energy Laboratory; and the author.

REFERENCES

1. Chester L. Cooper and William Pennell, Sustainable Development: From Concept to Implementation. Draft Concept Paper, Available from the authors at Battelle Pacific Northwest Laboratory, Washington, D.C., July 1994.

2. Harvey Brooks, "Sustainability and technology," in Science and Sustainability, Selected Papers. For IIASA's 20th Anniversary, International Institute for Applied Systems Analysis, Laxenburg, Austria, 1992, pp. 1–31.

3. The Sustainable Seattle Indicators of Sustainable Community. Draft for Review and Comment, Sustainable Seattle, Seattle, WA, June 24, 1993.

4. Diana M. Liverman, Mark E. Hanson, Becky J. Brown and Robert W. Meredith, Jr., "Forum—global sustainability: toward measurement," *Environ. Manage.,* 12(2):133–143, 1988.

5. Albert Adriaanse, Environmental Policy Performance Indicators: A Study on the Development of Indicators for Environmental Policy in the Netherlands, The Directorate for Information and External Relations of the Ministry of Housing, Physical Planning, and Environment, The Hague, The Netherlands, April 1993.

6. Interim Report of the Interagency Working Group on Sustainable Development Indicators. Draft Document for U.S. Government Review Only, July 6, 1994.

7. "Integrated economic and environmental satellite accounts" and "Accounting for mineral resources: issues and BEA's initial estimates," in Survey of Current Business, Department of Commerce, Bureau of Census, Washington, D.C., April 1994.

8. Lester R. Brown et al., *Vital Signs: 1994—The Trends That Are Shaping Our Future,* W.W. Norton & Company, New York, 1994, 160 pp.

6

AIR QUALITY, LAND USE AND TRANSPORTATION PERSPECTIVES: A CALIFORNIA CASE STUDY

Victoria Evans

THE SUSTAINABLE DEVELOPMENT CONCEPT

For development to be sustainable over the long term, physical and ecological sustainability are both preferable, if not critical. In our present operating system, we draw upon resources from the environment, many of which are nonrenewable. During extraction and use of these resources, we often discard contaminated residuals and wastes that are detrimental to our health as well

Victoria A. Evans is a Member Services Representative for the Environment Group at the Electric Power Research Institute, Palo Alto, California and has over twenty years of experience as an environmental regulatory analyst and air quality program manager focused in the energy sector. She has prepared EIR/EISs and regulatory strategy reports and performed permit feasibility analysis, technology assessment, siting analysis and compliance monitoring. Having worked in or for federal and state agencies, she has developed visibility standards and motor vehicle trip reduction programs. She has an M.S. in natural resource policy and administration and a B.S. in natural resource science from the University of Michigan.

©St. Lucie Press CCC 1-57444-079-9 1/97/$100/$.50

as to the environment. And we partially or sometimes completely diminish the earth's ability to regenerate. Since our existing economic system often undervalues the worth of the resource, a level of resource consumption is promoted that does not recognize the resource's value nor usually any future cleanup costs.[1]

This system does not provide for considering the economic value of the resource equitably with the cost of resource consumption. And the frequent occurrence of the lack of this balancing does not lead to decisions that support long-term physical and ecological sustainability. The focus of this chapter is on air quality and land use perspectives on sustainable development, with an emphasis upon interrelationships among air quality, land use and physical sustainability.

We may need to supplement existing policy analysis tools with new sustainability indicators that are either qualitative or fully quantitative indicators. This could include qualitative indicators such as avoidance of processes or products that create byproducts that are contaminated with toxic or persistent manmade substances (i.e., "negative legacies"). Examples of these are processes using CFCs or creating hazardous wastes as a manufacturing byproduct. Another example is the minimization and eventual avoidance of the high consumption rates of nonrenewable resources like fossil fuels. These two qualitative indicators are adapted from work by Swedish physician Dr. Robert on his framework for four ecological system conditions needed for sustainability.[2–4]

An example of a quantitative indicator is a resource efficiency quotient. A resource efficiency quotient could be based on the calculation of the output units of production divided by the natural resource input, multiplied by a factor reflecting the environmental residuals resulting from the process or product. The output factor of the calculation would be measured in kilowatts of electricity, barrels of oil, gallons of gasoline, number of refrigerators, tons of chemical fertilizers, etc. The quotient would be calculated relevant to an appropriate time period or usage factor (e.g., kilowatts per month, gallons per mile, barrels per gallon of product, etc.). Then highest resource efficiency quotient or "score" would occur for the process or product with the highest overall net resource efficiency and the lowest environmental residuals. Thus, the highest quotient would occur for the process or product that (1) created little to no persistent or toxic byproducts and (2) lowered consumption rates of renewable resources.

An existing example of a similar indicator being used as a surrogate for quantifying environmental impacts is the calculation of costs of externalities associated with alternative modes of electric generation. In some states, such a calculation is required of utilities by law as part of long-range planning for electric generation, known as Integrated Resource Planning.

Land is a fixed natural resource. It is not a renewable resource; however, it is reclaimable. Air is also a fixed natural resource, with the proportion of the gases and particles in the air changing over time and not always in the direction of supporting the long-term survival of life. The quality of ambient air is subject to the effects of meteorology, natural cleansing actions and atmospheric chemical reactions and conversions. In the atmosphere, long-range transport of air pollutants also occurs between an emissions source and a sensitive receptor. This is evidenced both by acid rain detection far from sources of air emissions and by the role of atmospheric CFCs in diminishing stratospheric ozone. In estimating the regenerative capacity of both air and land resources, we assume certain rates of impact which, due to improvements in technology and pollution prevention, may actually decrease gradually over time.

In California, due to the limited natural assimilative capacity of the atmosphere for air pollutants, ambient air quality problems are severe. The need for emissions reductions has transcended the scope of a solely technology-based solution. Thus, air quality agencies have had to reach beyond technological fixes. Concurrent with requiring reductions in stationary source emissions, the state of California has also required substantial reductions in motor vehicle emissions. California's land use patterns render residents predominately motor vehicle dependent. Improvements in vehicle combustion, the trapping of fuel evaporative emissions and improvements in gasoline quality have all significantly reduced vehicle emissions in California by using engineering solutions.

However, motor vehicle trip reduction measures have also been developed that require behavioral changes in drivers. Land use policies have also had to be developed to try to encourage fewer motor vehicle trips, more use of transit, and bicycle and pedestrian trips. The unique air quality problem created by the limited natural assimilative capacity of the atmosphere in California has fostered an awareness of the issue of sustainability in the clash between a motor-vehicle-dependent economy/lifestyle and healthful air quality.

FUTURE POLICY NEEDS FOR SUSTAINABILITY

The use of traditional tools for natural resource management is possible when ambient conditions do not change much or very quickly. The problem is that unprecedented rates of change are expected. And these changes are expected across a broad spectrum—global climate, population of people and cars, and technology. Little is known about how to design long-term environmental policies when significant uncertainty is expected. Thus, policies will need to allow for change over time.

I suggest that for addressing sustainability we need long-term policies that strengthen the resilience and integrity of natural and human-made systems. Sustainability requires more responsible environmental decision making today, made with a commitment to a view toward the long term. To incorporate sustainability will require decisions to be made based on an equitable accounting of all costs borne today and in the future. Decision making will need to include consideration of "negative legacies" or environmental residuals. Integration of a restructured cost-accounting approach and sustainable planning and decision-making guidelines will be needed. To be successful, this approach will need to be applied across sectors, both public and private.

The following are a set of proposed actions for sustainable development policy:

(1) Set quality of life levels

(2) Develop sustainable development indicators

(3) Evaluate the physical impact on sustainability of alternative policies and use the results during policy decision making

(4) Review the impacts/effectiveness of policies implemented and adapt them periodically as needed

Two key attributes suggested for these types of policies include (1) policy setting that does not foreclose other options and (2) actions to implement them that undergo timely review and adaptation over time. This latter element is very important, i.e., to include encouragement of continuing, timely evaluation and review of policies and implementing regulations and adaptation as needed over time.

POLICY IMPLEMENTATION FOR SUSTAINABILITY

To be successful, efforts toward sustainability appear to be needed at several levels of implementation and across multiple institutional/jurisdictional boundaries. These multiple levels of implementation include policy analysis, program development, program management, monitoring and measurement, technology transfer and pollution prevention. The scope of needed environmental problem solving does not always follow jurisdictional boundaries. For example, county and city boundaries do not always follow natural boundaries such as watersheds and airsheds.

Thus, decision making will be needed that cuts across the natural boundaries, jurisdictional boundaries and interconnected sectors (energy, waste management, air quality). This is expected to be very difficult.[5] Recent experience with vehicle trip reduction in California (described below) has demonstrated the difficulty and controversial nature of these issues.

Refinement, not reinvention, of existing institutional, regulatory and market processes is proposed here to incorporate sustainability issues. There is a substantial platform of existing environmental and planning policy processes for incorporating sustainability issues. These policy processes include, for example, land use planning; federal, state and local Environmental Impact Statement and Report requirements; air quality management; solid waste management; and electricity planning/regulation.

Some examples of market-based incentives and disincentives already in place that reduce or avoid resource demand include energy efficiency standards, time-of-day electric rates and demand-side management for both water and electricity. Measures that discourage production of environmentally harmful residuals include air and hazardous waste regulations that are pollution prevention based and requirements for use of recycled products. Another specific measure offers the multiple benefits of energy conservation, air pollution reduction and global climate change. This measure is the planting of street shade trees. The effects are cooling and insulating for buildings, cooling parked petroleum-fueled vehicles to reduce emissions from the evaporation of gasoline and also increasing plant respiration which adds oxygen to the atmosphere.

For "pollution prevention" from motor vehicle trips, demand-side management, or transportation system demand management, can be considered. Transportation control measures (TCMs) seek to reduce motor vehicle trips, particularly those taken by a single occupant to a single destination and return. For reducing motor vehicle trips, this can mean encouraging mixed land use developments with services and employment within walking distance.

What is important is not necessarily the cumulative number of homes or cars in a region but the amount of resource consumption per home and per capita. Thus, "technological fixes" (i.e., advancements in gasoline efficiency and pollutant reduction in motor vehicles) are important, but so are behavioral changes in motor vehicle trip-making. We will also need innovative methods to reengineer processes to be more sustainable and to measure the effects of these types of policies upon our objective of sustainability.

LAND USE AND AIR QUALITY

Probably the dominant factor influencing land use in the United States from World War II to the present is the personal passenger petroleum-fueled vehicle. We would not care so much if these vehicles were fueled with something that did not emit air pollutants, but they do. We might not still care so much, except that these vehicles are the dominant mode of transportation in the United States and in many parts of the world. Ironically, the use of petroleum-fueled motor vehicles has implications not only for air pollution but also for energy consumption and congestion, as well as for other sustainable development issues (land consumption).

By some estimates, as much as 50% of the land in urban communities is covered with concrete and asphalt needed to serve the automobile for parking lots, streets, freeways, car sales lots and gas stations. Sprawl development not only does not support mass transit, but also consumes land needed for growing food, preserving animal and plant species, recharging groundwater tables and for recreation. More compact development and mixed-use development reduces costs of public services (police, fire, sanitation and transportation) by 40 to 400%, in addition to reducing the need to make multiple car trips.[6]

In most western U.S. cities that do not attain the federal ambient air quality standards, from one-half to three-quarters of the ozone and carbon monoxide (CO) are attributable to motor vehicles. The operation of on-road motor vehicles and marketing of gasoline in most of the large urban areas in the United States contribute over half of the total emissions of CO, hydrocarbons (HC) and nitrogen oxides (NO_x). Precursors of photochemical oxidants include emissions of HC and NO_x. Levels of CO and photochemical oxidants (primarily ozone) in ambient air are above federal and state health standards.[7]

Under recent legislative mandates, government agencies have developed programs to reduce emissions of CO and of ozone precursors. Table 6.1 provides a summary of these types of programs. These air pollution control measures include engineering emissions controls such as tailpipe emissions and gasoline quality and also nonengineering-based controls or transportation control measures such as motor vehicle trip reduction, reduction of motor vehicle miles traveled and indirect source measures. A major type of indirect sources is land uses that attract or generate motor vehicle trips (shopping malls, housing developments, airports and medical centers).

What about land use strategies that are better for at least our physical environment, including air quality? Under the federal 1990 Clean Air Act Amendments, the degree of state and local control required for mobile sources depends on the area's degree of nonattainment of both the ozone and CO standards. Three types of controls are required of states for mobile sources under the 1990 amendments: enhanced smog check programs, use of gasoline vapor control equipment at gas stations and TCMs. Only the latter types of measures involve land use planning. The Clean Air Amendments include transportation performance standards that require implementation of enforceable, reasonably available TCMs that offset emissions due to growth in vehicle miles traveled (VMT) and trips, in the areas with the worst ozone and CO levels.

California also has mandates for local air districts to control both indirect (land use) and mobile sources. Examples from

TABLE 6.1 Motor Vehicle Programs to Reduce Emissions of Ozone Precursor and CO in the United States[7]

Transportation control measures		Emissions controls/reduction	
Trip and VMT reduction	*Vehicle transportation improvements*	*Vehicle tailpipe*	*Fuels*
Employer trip reduction	Signals timed to lessen stop-and-go on arterials	New car emission standards	Vapor recovery from gasoline
City trip reduction ordinances		Vehicle inspection and maintenance	Oxygenated gasoline
Light rail/transit		On-board vapor recovery	Vapor pressure reduction
Indirect source review		Early retirement of old vehicles (buy-back programs)	Reformulated gasoline
Bicycle lanes, pedestrian paths/access			Ultra-low sulfur gasoline
High Occupancy Vehicle (HOV) lanes			Clean fuel vehicles (CNG, propane, electric, etc.)
Parking lot restrictions			

California will be discussed due to the author's experience with these air programs. Since a substantial amount of suburban-based development has occurred in California, the California Clean Air Act mandated performance standards for determining the sufficiency and measurability of the effectiveness of TCMs. Levels of Average Vehicle Occupancy were legislated as performance standards by the severity of the area's ozone levels. Also, a requirement for no net increase in vehicle emissions after 1997 was mandated for the most serious ozone areas. Credit is allowed for measures that concurrently increase persons per vehicle and reduce vehicle trips. Also, credit is given for alternative strategies that achieve emissions reductions, such as old car buyback programs.[8–10]

TRANSPORTATION AND LAND USE

TCMs are intended to reduce emissions from mobile sources through measures such as employer trip reduction, trip reduction ordinances and transit and bike path improvement programs. The focus is on vehicle trips since almost three-quarters of the air pollutants are produced just from starting a cold engine for a short five-mile trip. The predominant number of vehicle trips are short—between five to ten miles. However, the number of miles traveled by cars has been growing at two times the rate of the population. As the number of cars increases, traffic becomes congested, and as driving time doubles, hydrocarbon emissions almost triple.[9]

TCMs such as employer trip reduction rules, parking management ordinances and restrictions on vehicle operations fall within the authority of air districts and cities and counties. Long-range land development policies that support reductions in vehicle trips are also included but must be accomplished in the context of a regionwide transportation plan and local general land use plans.

Employer Trip Reduction

Agencies managing air quality in three of the major urban areas in California—Los Angeles, San Joaquin Valley and San Francisco—have adopted employer trip reduction (ETR) rules. All of the ETR rules include only trip reduction provisions and do not affect land use directly. All three rules affect employers of over one hundred employees and require a transportation coordinator and a company plan to reach an average vehicle ridership target. The requirements affect public agencies as well as private companies. The Los Angeles area air agency first adopted its ETR rule in 1987; it affects over ten million people in four counties and over eighty cities. The San Francisco area adopted its ETR rule in late 1992, affecting almost four million people in eight counties and seventy-five cities. The San Joaquin Valley adopted its ETR rule in early 1994. Current California requirements for ozone nonattainment areas are for an ETR rule that calls for an average of 1.5 persons per passenger vehicle during commute hours by 1999.[7,11]

Trip Reduction Ordinances

An overwhelming number of counties and cities in California have adopted trip reduction ordinances (TROs) to increase use of mass transit, carpooling and telecommuting for commuters. Some of these measures also restrict the amount of parking for new development. The vast majority of TROs adopted in California cities and counties are employer-based TROs. Almost all of the affected California cities with TROs in place that were adopted before the ETR rules will have to substantially amend them to receive delegation from each local air agency to administer their more aggressive ETR programs.

TROs apply to both existing and future employers based upon the number of employees. TRO requirements can also be triggered based upon employees per square foot, land use type and gross square feet. Several local governments have adopted broader-based programs that affect new developments and affect expansion or changes in existing land uses.

Indirect Source Review Programs

U.S. areas that do not attain the federal ozone standard *may* adopt provisions for an indirect source program. Similar areas in California must adopt a program. Emissions associated with indirect sources are primarily related to land uses attracting trips in on-road motor vehicles. However, emissions from area sources are also included in this category. Area sources include such emission sources as gas stations and residential natural gas heaters. Indirect source programs can include policies, procedures or regulatory requirements affecting land use decisions, in addition to TCMs. There is a close relationship between the requirements for both indirect source programs and TCMs.[12]

California allows TCMs to be part of an indirect source program, but the U.S. EPA does not. Through their police power, local governments can use ordinances, environmental impact review, discretionary permit processes, capital improvement programs, business licenses and other means to reduce vehicle trips. Other strategies to reduce indirect emissions include nonmotorized transportation and transit, parking management, energy conservation (to lower natural gas and electricity consumption), low-emitting solvents and coatings (building materials) and solid waste management (collection routes, transfer and processing facilities and recycling).

The three major California areas discussed above also proposed that air quality issues be more fully addressed in reviews under the California Environmental Quality Act (CEQA) or the state environmental impact assessment process. These air agencies will provide technical assistance for assessing impacts; identifying appropriate mitigation, modeling or data needs; and monitoring to determine mitigation strategy effectiveness.[12,13]

Also, the Los Angeles area and San Joaquin Valley air agencies provided technical guidance and encouragement to local cities and counties to adopt air quality policies for local land use planning. While state authority is given to controlling emissions due to indirect sources, current California law does not permit air agencies to control land use. Therefore, air agencies are proposing to influence the reduction of mobile source emissions by affecting transportation patterns, such as the location of new housing, employment and transportation facilities, through the CEQA review process and through local city or county General Plan Elements or Air Quality Programs.[12]

Tools for estimating emissions from new or modified indirect sources for these purposes and for estimating TCM effectiveness have been recently developed.[14,15] Relatively simple methods for estimating both construction and operational emissions from a variety of indirect sources were developed for the Los Angeles area air agency and published in its CEQA Handbook.[13,16,17]

The air quality issues highlighted in "Air Quality Elements" can include both urban growth and residential wood combustion. The goals can include ensuring that:[12]

- Local land use decisions consider air quality implications

- Local land use patterns reduce vehicle trips and travel distances

- Orderly growth and expansion reduce vehicle miles traveled

- Safe and convenient bicycle and pedestrian networks and support facilities be provided

- Coordinated and efficient bus services provide an effective alternative to auto use

- Measures reduce traffic congestion and associated emissions

- Energy efficiency in new building development will reduce the need for electric generation

- Other programs reduce single-passenger vehicle use and resulting emissions

An example of the policies and programs that begin to address the sustainability of land use and air quality is discussed below.

MOTOR VEHICLE TRIP REDUCTION IN SOUTHERN CALIFORNIA

Transportation Control Measures

In Southern California, local governments were called upon to assist in implementing the South Coast Air Quality Management District's plans for attainment of the federal standards for ozone and carbon monoxide. Plans call for action by local governments to adopt and implement transportation demand measures (TDMs) and growth management measures to reduce vehicle trips (VTs) and VMT within the four counties of the Los Angeles area air basin. The area has the worst air quality in the nation, with a population of over ten million people and eight million cars. (TDMs and growth management measures will be referred to together as transportation control measures [TCMs].)

These TCMs address both work and nonwork trips and include measures for nonmotorized transportation, alternative work weeks, delivery services, shuttles (to and from transit stops), densification, mixed land use, lunchtime shuttles, park-and-ride lots, parking management, facility requirements for carpools/vanpools, transit improvements and parking pricing. These TCMs are to be incorporated into local governmental policy, as part of a local ordinance or specific land use plans in the over eighty-five cities and four counties within the South Coast Air Basin.

A strategy and a menu of individual TCM "actions" were developed to respond to local governments' requests for guidance on what to do, by when and specific methods for achieving the emissions reduction targets.[18–21] In addition, methods for implementation and quantification of the TCM actions were developed, since the California Air Resources Board and the U.S. EPA were concerned as to whether the TCMs would be enforceable and quantifiable.

(1) **Menu-based approach**—The menu of actions provides a wide range of alternative individual TCM actions or measures that can be utilized in a local strategy to reduce VT or VMT. Local government planners screen the menu of potential actions, select those applicable and estimate whether the actions could achieve their trip reduction target.

(2) **Regional performance goals**—Regional performance goals for trip reduction targets (in VT) are allocated as proportional shares for each county and then disaggregated for each city. Local jurisdictions are to develop their own programs to meet these overall targets. By combining all of the individual trip reductions into one performance target, local governments are afforded the flexibility to select a combination of actions most suited to them to meet their target. Thus, the program is performance based rather than prescriptive for compliance.

(3) **Simple methods for quantification of effectiveness**—Methods to quantify TCM effectiveness based on past implementation of each action to reduce VT were provided. These methods gauge the relative value to local jurisdictions to meet their targets. Also, to meet state and federal requirements for quantification of progress in comparison to

mandated transportation performance standards, this performance-based approach was developed. For each TCM action, an upper limit on the expected efficiency allowed for the action was provided. This ceiling limit on effectiveness, along with recommendations for packaging TCMs to take advantage of synergistic effects and to avoid conflicting effects, attempts to address the interactive impacts of the application of multiple TCMs.

(4) **Single performance measure, vehicle trips**—To simplify monitoring and administration, one indicator for the program was selected: reduced VT. The responsibility for converting VT and VMT reduced to motor vehicle emissions reduced is with the Air District technical staff. This approach was selected due to the controversy over the underestimation of motor vehicle emissions.[22] Whether the action is a TDM, TCM, indirect source or growth management type measure, the results of all of these measures can be translated to VT reductions.

For actions that decrease VMT, an equivalency factor, an emissions equivalent vehicle trip (EEVT), was developed to convert VMT to VT. The factor is based on county-specific average trip lengths and the relationship of emissions between VT and VMT.[20] However, since VT elimination generates higher emission reductions than TCMs that reduce VMT, actions that directly reduce VT will more readily help a local jurisdiction meet its market share VT reduction target. The EEVT methodology mathematically converts VMT reductions to a VT equivalent.

Implementation

For each jurisdiction, the types of trip (e.g., trip destinations that could be affected by each action versus the baseline number of trips) are evaluated in a screening-level analysis. Next, a planning-level analysis method is used to calculate VT and VMT reductions from each trip reduction action specifically. These simple planning analysis methods were developed to balance the need for technical integrity comparable to detailed TCM models and the need to maintain simplicity for local government planners.

The air quality benefits of these measures are calculated based upon the effectiveness of each measure upon the reduction of either VT or VMT. VMT credit is adjusted to an equivalent VT utilizing the factor (described above) that properly accounts for associated air emissions. Methods are included for calculating VT reductions for work and for all trips from bicycle improvements and pedestrian improvements and also for work trips from telecommuting, alternative work weeks, ridesharing and rideshare support facilities. The lack of information on the benefits of densification and mixed land use prevented quantification method development and is indicative of the paucity of historical data on VT and VMT reductions experienced in the application of these types of growth management measures.

CONCLUSIONS AND RECOMMENDATIONS

While state authority is given to controlling emissions due to indirect sources, current law does not permit air agencies to control land use. Air agencies cannot dictate, but only recommend, where new housing, employment and transportation facilities might be located to minimize VT and VMT.[12] Therefore, air agencies are proposing to address the reduction of mobile source emissions by affecting transportation patterns, such as the location of new housing, employment and transportation facilities, through the environmental review process and through General Plan Elements or Air Quality Programs.

Air districts appear to recognize the limitations to their indirect authority in controlling land use decisions and that too active a role on their part might potentially create conflicts with local government over jurisdiction in land use decisions. As demonstrated by indirect source control efforts taken to date, the Los Angeles area agency (reported on here) and three other major California districts (reported on elsewhere) prefer that local jurisdictions take on the primary responsibility for indirect source emissions reduction.[11,12]

Air agencies appear to prefer that local governments adopt programs through local TROs or through incorporating air quality goals in planning, development and land use decision making. All air agencies appear to be primarily concerned that local governments be sincerely committed to adequately fulfill these responsibilities and to provide a demonstration of the effectiveness of these local programs. This is the dilemma posed by multiple levels of jurisdictional boundaries and the need to address interconnected sectors, air quality and land use, to develop effective sustainable policies.

Monitoring of these types of land use/air quality programs is critical to develop a better data base on the effectiveness of measures that increase sustainability. For example, future refinements of the estimated levels of trip reduction and quantification methods are dependent upon adequate data collection and analysis during program monitoring. Methods for quantifying the VT and VMT benefits of measures like densification and mixed land use that particularly reduce VT are especially important, since these measures are likely to have the most significant and long-term effects.

Any study should examine the enforceability, efficiency, effectiveness, cost effectiveness, ease of compliance, methods for calculating emission reductions and any institutional or other barriers impeding effectiveness. Special questions should be included to inquire about equity issues and to identify transportation needs of low-income workers. Additional work should also focus upon collecting information and refining methods to quantify the synergistic effects of TCMs as experienced in practice.

In the short run, these types of programs focus upon mode shift of trips (e.g., fewer single-occupant vehicles, fewer motor vehicle trips and slowing the growth of VMT). Future emphasis should continue to be placed upon time-of-day (TOD) considerations in affecting trip-making by trip types. The major benefit of TOD-based policies is that they emphasize the use of existing system capacity or infrastructure. Thus, TOD policies are included as TCMs in California attainment plans in an attempt to shift motor vehicle trips off-peak (out of typical commuter hours) to reduce traffic congestion and air emissions. Vehicle air emissions (and gasoline consumption) are higher during lower vehicle speeds, stop-and-go traffic and idling. Reducing vehicle combustion emissions at their peak hours of generation will, therefore, decrease the peak ambient levels of ozone, CO and other pollutants.

Current programs affect the TOD of trips through the timing of high-occupancy-lane use restrictions and use of ramp metering. Developing measures to further influence trip-making by TOD are cost effective, since they encourage the use of existing highway capacity. Future TCM development based on TOD considerations could benefit from a review of the policies and framework used in developing similar policies for TOD electricity pricing; in the mid-1970 policies were devised to encourage off-peak use of electric generating capacity. Design of measures and supporting data collection to encourage off-peak use of the existing transportation infrastructure should receive more attention.[11]

Finally, there is general agreement that existing mobile source emissions, and therefore emissions inventories, are underestimated by a factor of between 2 and 3.[22] Considering this underestimation, the net benefits of emissions reductions from VT and VMT reductions will be greater than that predicted. Nonetheless, there is still an issue of concern related to planning of emissions reductions needed from motor vehicles.

The proportion of the emissions inventory attributable to motor vehicles is used to generally determine the level of emissions reductions needed from transportation and growth management measures. As this proportion of the inventory is increased, air agencies will want to review the level of emissions reductions in the Air Quality Management Plan that are appropriate from the motor vehicle category. This review may result in the need to revise the level of VT reductions that is needed from trip reduction and growth management programs.

REFERENCES

1. California Energy Commission, Energy-Aware Planning Guide, January 1993.

2. K.H. Robert, *Non-Negotiable Facts as a Basis for Decision-Making,* The Natural Step Foundation, Sweden.

3. K.H. Robert, "Educating a nation: the natural step," *In Context: A Quarterly of Humane Sustainable Culture,* 1993.

4. R. Eronn, "The natural step—a social invention for the environment," *Current,* No. 401, Dec. 1993.

5. K. Sessions, "Building the capacity for change," *EPA Journal,* 19(2): 15–19, 1993.

6. Local Government Commission, Land Use Strategies for More Livable Places, May 1992.

7. P.B. Bosserman and V.A. Evans, "Loving our cars, exhausting clean air—a survey of programs to abate CO, HC and NO_x from motor vehicles on the western coast of North America," presented at the International Symposium on Transport and Air Pollution, Avignon, France, June 6–10, 1994.

8. California Air Resources Board, California Clean Air Act Guidance for the Development of Indirect Source Control Programs, July 1990.

9. California Air Resources Board, California Clean Air Act Guidance for the Development of Indirect Source Control Programs. Technical Support Document, July 1990.

10. California Air Resources Board, Office of Strategic Planning, The Air Pollution–Transportation Linkage, 1989, 10 pp.

11. V. Evans and R. Jones, "A survey of trip reduction ordinances and transportation control measures in California air districts," in Proceedings 86th Annual Meeting of the Air and Waste Management Association, Denver, CO, June 1993.

12. V. Evans and R. Jones, "An overview of indirect source control programs in ozone nonattainment plans in California," in Proceedings 86th Annual Meeting of the Air and Waste Management Association, Denver, CO, June 1993.

13. V.A. Evans and C. Day, "Experience with a program to reduce emissions of indirect sources in Southern California," in Proceedings Air and Waste Management Association 87th Annual Meeting, Cincinnati, OH, June 19–24, 1994.

14. Sierra Research, Inc. and JHK & Associates, Methodologies for Quantifying the Emission Reductions of Transportation Control Measures. Prepared for San Diego Association of Governments, 1991.

15. B.S. Austin, J.G. Duvall, D.S. Eisinger, J.G. Heiken and S.B. Shepard, Estimating Travel and Emission Effects of TCMs. Prepared for U.S. Environmental Protection Agency, Office of Mobile Sources, by Systems Applications International, 1991.

16. T. Dodson and V. Evans, Draft Air Quality Handbook for Implementing the California Environmental Quality Act. Prepared for South Coast Air Quality Management District by Tom Dodson and Associates and Gaia Associates, 1990, 292 pp.

17. South Coast Air Quality Management District, Final Air Quality Handbook for Implementing the California Environmental Quality Act, 1993.

18. V. Evans, R. Jones and D. Morrow, Development of Trip Reduction and Growth Management Measures and Quantification of Associated Reductions in Motor Vehicle Trips and Vehicle Miles Traveled in the South Coast Air Basin. Draft report prepared for the South Coast Air Quality Management District, 1992.

19. V. Evans and D. Morrow, "Development of a program framework and implementation process for the transportation control and growth management program in the South Coast Air Quality Management District," in Proceedings 86th Annual Meeting of the Air and Waste Management Association, Denver, CO, June 1993.

20. V. Evans and D. Morrow, "Simple methodologies for quantifying VT and VMT reductions from transportation control measures," in Proceedings 86th Annual Meeting of the Air and Waste Management Association, Denver, CO, June 1993.

21. South Coast Air Quality Management District, Handbook for Preparing a Local Government Trip Reduction Ordinance, 1993.

22. E.M. Fujita, B.E. Croes, C.L. Bennett, D.R. Lawson, F.W. Lurmann and H.H. Main, "Comparison of emission inventory and ambient concentration ratios of CO, NMOG, and NO_x in California's South Coast Air Basin," *J. Air Waste Manage. Assoc.*, 42:264–276, 1992.

BIBLIOGRAPHY

Global Environmental Management Initiative (GEMI), Environmental Self-Assessment Program, 1992, 111 pp.

P. Hawken, "A declaration of sustainability," *Utne Reader,* Sept./Oct. 1993.

P. Hawken, *The Ecology of Commerce,* Harper-Collins, New York, 1993, 250 pp.

"Resource efficiency is the key to sustainable development," *Environ. Bus. J.,* VII(7), July 1994.

San Joaquin Valley Unified Air Pollution Control District, Final Environmental Impact Report for the San Joaquin Valley and 1991 California Clean Air Act Air Quality Attainment Plan, August 1991.

South Coast Air Quality Management District, Making Clean Air a Priority, A Guide for Planners in Local Governments, September 1990.

INDEX